Mathematical Approaches to Software Quality

Gerard O'Regan

Mathematical Approaches to Software Quality

With 52 Figures

 Springer

Gerard O'Regan, BSc, MSc, PhD
11 White Oaks, Mallow, Co. Cork, Ireland
oregang@yahoo.com

British Library Cataloguing in Publication Data
A catalogue record for this book is available from the British Library

Library of Congress Control Number: 2005935914

ISBN-10: 1-84628-242-X   Printed on acid-free paper
ISBN-13: 978-1-84628-242-3

Printed in the United States of America   (SPI/EB)

9 8 7 6 5 4 3 2 1

Springer Science+Business Media
springer.com

To the memory of Con and Eily O'Regan
(Too dearly loved to be forgotten)

Preface

Overview

The objective of this book is to provide a comprehensive introduction to mathematical approaches that can assist in achieving high-quality software. An introduction to mathematics that is essential for sound software engineering is provided, as well as a discussion of the various mathematical methods that are used in academia and (to varying degrees) in industry. The mathematical approaches considered here include the Z specification language; the Vienna Development Method (VDM); the Irish school of VDM (VDM$^{\bullet}$); the axiomatic approach of Dijkstra and Hoare; the classical engineering approach of Parnas; the Cleanroom approach developed at IBM; software reliability, and the unified modeling language (UML). Finally, the challenge of technology transfer of the mathematical methods to industry is considered.

The book aims to explain the main features of the mathematical approaches to the reader, and to thereby assist the reader in applying these methods to solve practical problems. The chapter on technology transfer presents an overview of how these technologies may be transferred to industry, and includes a discussion on the usability of formal methods and pilots of formal methods.

Organization and Features

Chapter 1 provides an introduction to an engineering approach to software development using mathematical techniques. A review of the more popular formal methods is presented. These include the model-oriented approaches of Z or VDM, and the axiomatic approaches such as the Communicating Sequential Processes (CSP). The nature and role of mathematical proof is discussed, as mathematical proof has the potential to play a key role in program verification.

The second chapter considers the mathematics required for sound software engineering. This includes discrete mathematics such as set theory, functions and relations; propositional and predicate logic for software engineers; tabular expressions as developed by Parnas; probability and applied statistics for the prediction of software reliability; calculus and matrix theory; finite state machines; and graph theory.

Chapter 3 is a detailed examination of mathematical logic including propositional and predicate calculus, as well as an examination of ways to deal with undefined values that arise in specification. The three approaches to undefinedness considered are the logic of partial functions (LPFs) developed by Cliff Jones; the approach of Parnas that essentially treats a primitive predicate calculus expression containing an undefined value as false, and thereby preserving a 2-valued logic; and the approach of Dijkstra that uses the **cand** and **cor** operators.

The next three chapters are concerned with the model-oriented approach of formal specification. Chapter 4 on Z includes the main features of the specification language as well as the schema calculus. Chapter 5 on VDM describes the history of its development at IBM in Vienna, the main features of the language and its development method. Chapter 6 on VDM* explains the philosophy of the Irish school of VDM, and explains how it differs from standard VDM. Z and VDM are the two most widely used formal specification languages and have been employed in a number of industrial projects.

Chapter 7 focuses on the approach of Dijkstra and Hoare including the calculus of weakest preconditions developed by Dijkstra and the axiomatic semantics of programming languages developed by Hoare. Chapter 8 discusses the classical engineering approach of Parnas, and includes a discussion on tabular expressions. Tabular expressions have been employed to provide a mathematical specification of the requirements of the A-7 aircraft and to certify the shutdown software for the Darlington nuclear power plant.

Chapter 9 is concerned with the Cleanroom approach of Harlan Mills and the mathematics of software reliability. Cleanroom enables a mathematical prediction of the software reliability to be made based on the expected usage of the software. The software reliability is expressed in terms of the mean time to failure (MTTF). Chapter 10 is concerned with the unified modeling language (UML). This is a visual approach to the formal specification and design of software, and the UML diagrams provide different viewpoints of the proposed system. The final chapter is concerned with the technology transfer of the mathematical approaches to industry. This includes piloting formal methods on one or more projects to evaluate their suitability for the organization.

Audience

This book is suitable for undergraduate and graduate computer science students who are interested in an overview of mathematical methods that may be employed to develop high-quality software. The material is mathematical but is presented as simply as possible. Motivated software and quality engineers who are interested in knowing how a mathematical approach can assist in achieving high-quality software will find the book to be a useful overview.

Acknowledgments

I am deeply indebted to friends and colleagues in industry and academia who supported my efforts in this endeavor. I would like to thank the team at Springer for their professional work and a special thanks to the copyeditor (M. Bearden) for his many helpful comments and suggestions. Finally, I would also like to thank family and personal friends such as Kevin and Maura for support.

Gerard O'Regan.
November 2005

Contents:

1
Introduction

NATO organized two famous conferences on software engineering in the late 1960s. The first conference was held in Garmisch, Germany, in 1968 and was followed by a second conference in Rome in 1969. The NATO conferences highlighted the problems that existed in the software sector in the late 1960s and the term *software crisis* was coined to refer to the problems associated with software projects. These included budget and schedule overruns and problems with the quality and reliability of the delivered software. This led to the birth of *software engineering* as a separate discipline and the realization that programming is quite distinct from science and mathematics. Programmers are like engineers in the sense that they build products; however, programmers are not educated as traditional engineers as they receive minimal education in design and mathematics. Consequently, Parnas argues [Par:01] that problems with software can be expected if individuals are neither properly educated nor qualified for the job that they are performing.[1]

The construction of bridges was problematic in the 19th century and many people who presented themselves as qualified to design and construct bridges did not have the required knowledge and expertise. Consequently, many bridges collapsed, endangering the lives of the public. This led to legislation requiring an engineer to be licensed by the Professional Engineering Association prior to practicing as an engineer. These associations identify a core body of knowledge that the engineer is required to possess and the licensing body verifies that the engineer has the required qualifications and experience. The licensing of engineers by most branches of engineering ensures that only personnel competent to design and build products actually do so. This in turn leads to products that the public can safely use; in other words, the engineer has a responsibility to ensure that the products are properly built and are safe to use.

---

[1] Software Companies that are following approaches such as the CMM or ISO 9000:2000 consider the qualification of staff before assigning staff to performing specific tasks. The approach adopted is that only staff with the appropriate qualifications and experience are assigned to the particular role. However, as Parnas has noted there is no professional software engineering body that licenses software engineers. My experience is that the more mature companies place significant emphasis on the education and continuous development of their staff and in introducing best practice in software engineering into their organization. I have observed an increasing trend among companies to mature their software processes to enable them to deliver superior results. One of the purposes that the original CMM served was to enable the U.S. Department of Defense (DOD) to have a mechanism to assess the capability and maturity of software subcontractors.

This is to be contrasted with the software engineering discipline where there is no licensing mechanism and where individuals with no qualifications can participate in building software products. Parnas argues that the resulting problems with software quality should come as no surprise as the work is being performed by staff who do not have the right education and training to perform their roles.[2]

The Standish group conducted research [ORg:02] on the extent of current problems with schedule and budget overruns of IT projects. This study was conducted in the United States but there is no reason to believe that European or Asian companies perform any better. The results indicate that 33% of projects are between 21 and 51% over estimate; 18% are between 51% and 100% over estimate; and 11% are between 101% and 200% over estimate.[3] Fred Brooks argues that software is inherently complex and that there is no silver bullet that will resolve all of the problems associated with software such as schedule overruns and software quality problems [Brk:75, Brk:86].

The problem with poor software quality and poor software design is very evident in the numerous security flaws exhibited in Microsoft Windows software. Such flaws can cause minor irritation at best or at worse can seriously disrupt the work of an organization or individual. The Y2K problem, where dates were represented in a 2-digit format, required major rework for year 2000 compliance. Clearly, well-designed programs would have hidden the representation of the date and would have required minimal changes for year 2000 compliance. However, the quality of software produced by some companies is superior.[4] These companies employ mature software processes and are committed to continuous improvement.

The focus of this book is on mathematical techniques that can assist software quality rather than on the problem of estimation and on-time delivery that is part of project management. Mathematics plays a key role in engineering and can assist software engineers in delivery of high-quality products that are safe to use. There is a lot of industrial interest in software process maturity for software organizations, and approaches to assess and mature software companies are described in [ORg:02].[5] These focus mainly on improving the effectiveness of the

---

[2] Parnas's comments are overly severe in my experience. The mature software companies do consider qualification of staff and employ rigorous processes including software inspections and testing to assure quality.

[3] It should be noted that these are IT projects covering diverse sectors including banking, telecommunications, etc., rather than pure software companies. My experience is that mature software companies following maturity frameworks such as level 2 or level 3 CMM maturity achieve project delivery generally within 20% of the project estimate. Mathematical approaches to software quality are focused on technical ways to achieve software quality. There is also the need to focus on the management side of software engineering also as this is essential for project success. The reader is referred to [ORg:02].

[4] I am familiar with projects at Motorola that achieved 5.6σ-quality in a L4 CMM environment (i.e., approx 20 defects per million lines of code, which represents very high quality).

[5] Approaches such as the CMM or SPICE (ISO 15504) focus mainly on the management and organizational practices required in software engineering. The emphasis is on defining and following the software process. However, there is insufficient technical detail on requirements, design, coding and

management and organization practices related to software engineering. The use of mathematical techniques is complementary to a maturity approach such as the Capability Maturity Model (CMM). The CMM allows the mathematical techniques to be properly piloted to verify that the approach suits the company and that superior results will be achieved by using mathematical methods. The bottom line for a software company is that the use of formal methods should provide a cost-effective solution. The CMM also includes practices to assist with a sound deployment of a new process.

1.1 Software Engineering

Software engineering involves multiperson construction of multiversion programs. Sound software engineering requires the engineer to state precisely the requirements that the software product is to satisfy and then to produce designs that will meet these requirements. Software engineers should start with a precise description of the problem to be solved; then proceed to producing a design and validating the correctness of the design; finally, implementation and testing are performed. The software requirements should clearly state what is required and it should be evident what is not required. Engineers are required to produce the product design and then analyze their design for correctness. An engineering analysis of the design includes mathematics and software inspections, and this is essential to ensure the correctness of the design. The term *engineer* is generally applied only to people who have attained the necessary education and competence to be called engineers and who base their practice on mathematical and scientific principles. Often in computer science the term engineer is employed loosely to refer to anyone who builds things rather than to an individual with a core set of knowledge, experience, and competence.

Parnas is a strong advocate of a classical engineering approach and argues that computer scientists should have the right education to apply scientific and mathematical principles in their work. Software engineers need to receive an appropriate education in mathematics and design in order to be able to build high-quality and safe products. Often, computer science courses tend to emphasize the latest programming language rather than the essential engineering foundations. The material learned in a computer science course is often outdated shortly after the student graduates. The problem is that students are generally taught programming and syntax but not how to design and analyse software. Computer science courses tend to include a small amount of mathematics whereas mathematics is a significant part of an engineering course. The engineering approach to the teaching of mathematics is to emphasize its application and especially the application to developing and analyzing product designs. The mathematics that software engineering students need to be taught includes sets, relations, functions, mathematical logic, tabular expression, and finite state

---

testing and the focus of this book is more on the mathematical approach needed for a good unambiguous definition of the requirements.

machines. The emphasis needs to be on the application of mathematics to solve practical problems.

The consequence of the existing gap in the education of current software engineers is that there is a need to retrain existing software engineers with a more solid engineering education.[6] Software engineers need education on specification, design, turning designs into programs, software inspections, and testing. The education should enable the software engineer to produce well-structured programs using module decomposition and information hiding.

Parnas has argued that software engineers have individual responsibilities as professionals.[7] They are responsible for designing and implementing high-quality and reliable software that is safe to use. They are also accountable for their own decisions and actions[8] and have a responsibility to object to decisions that violate professional standards. Professional engineers have a duty to their clients to ensure that they are solving the real problem of the client. They need to precisely state the problem before working on its solution. Engineers need to be honest about current capabilities when asked to work on problems that have no appropriate technical solution rather than accepting a contract for something that cannot be done.[9]

The licensing of a professional engineer provides confidence that the engineer has the right education and experience to build safe and reliable products. Otherwise, the profession gets a bad name as a result of poor work carried out by unqualified people. Professional engineers are required to follow rules of good practice and to object when rules are violated. The licensing of an engineer requires that the engineer completes an accepted engineering course and understands the professional responsibility of the engineer. The professional body is

---

[6] Realistically, a mass reeducation program for current software engineers is highly unlikely. Any retraining program for current software engineers would need to minimize the impact on company time as software companies are very focused on the bottom line in the very competitive environment of the new millennium. This includes intense competition from the low cost software development locations in Eastern Europe and Asia as outsourcing of software development to such locations is an increasing trend.

[7] The concept of accountability is not new; indeed the ancient Babylonians employed a code of laws c. 1750 B.C. known as the Hammarabi Code. This code included the law that if a house collapsed and killed the owner then the builder of the house would be executed.

[8] However, it is unlikely that an individual programmer would be subject to litigation in the case of a flaw in a program causing damage or loss of life. Most software products are accompanied by a comprehensive disclaimer of responsibility for problems rather than a guarantee of quality. Software engineering is a team-based activity involving many engineers in various parts of the project and it could be potentially difficult for an outside party to prove that the cause of a particular problem is due to the professional negligence of a particular software engineer, as there are many others involved in the process such as reviewers of documentation and code and the various test groups. Companies are more likely to be subject to litigation, as a company is legally responsible for the actions of their employees in the workplace. However, the legal aspects of licensing software may protect software companies from litigation including those companies that seem to place little emphasis on software quality. However, greater legal protection for the customer can be built into the contract between the supplier and the customer for bespoke-software development.

[9] Parnas applied this professional responsibility faithfully when he argued against the Strategic Defense Initiative (SDI) as he believed that the public (i.e., taxpayers) was being misled and that the goals of the project were not achievable.

responsible for enforcing standards and certification. The term engineer is a title that is awarded on merit but also places responsibilities on its holder.

Engineers have a professional responsibility and are required to behave ethically with their clients. The membership of the professional engineering body requires the member to adhere to the code of ethics of the profession. The code of ethics[10] will detail the ethical behavior and responsibilities including:

No.	Responsibility
1.	Honesty and fairness in dealings with clients.
2.	Responsibility for actions.
3.	Continuous learning to ensure appropriate knowledge to serve the client effectively.

Table 1.1. Professional Responsibilities of Software Engineers

The approach used in current software engineering is to follow a well-defined software engineering process. The process includes activities such as requirements gathering, requirements specification, architecture design, software design, coding,and testing. Most companies use a set of templates for the various phases (e.g., the IEEE standards). The waterfall model [Roy:70] and spiral model [Boe:88] are popular software development lifecycles.

The concept of process maturity has become popular with the Capability Maturity Model, and organizations such as the SEI have collected empirical data to suggest that there is a close relationship between software process maturity and the quality and the reliability of the delivered software. However, the main focus of the CMM is management and organization practices rather than on the technical engineering practices.

There has been a growth of popularity among software developers in light-weight methodologies such as XP [Bec:00]. These methodologies view documentation with distaste and often software development commences prior to the full specification of the requirements.

Classical engineering places emphasis on detailed planning and design and includes appropriate documentation. The design documents are analyzed and reviewed and used as a reference during the construction. Documentation is produced in the software sector with varying degrees of success. The popularity of a methodology such as XP suggests developer antipathy to documentation. However, the more mature companies recognize the value in software documentation and regard documentation as essential for all phases of the project. It is essential that the documentation is kept up to date and reflects the actual system since the impact of any changes requested to the software during maintenance cannot be properly determined. The empirical evidence suggests (e.g., [Let:03]) that software documentation is often out of date.

The difference between engineering documents and standard software documents (e.g., documentation following the IEEE templates) is that engineering

---

[10] My experience of working in software companies is that these are core values of most mature software companies. Larger companies have a code of ethics that employees are required to adhere to.

documents are precise enough to allow systematic analysis to be carried out, whereas since software documents employ natural language rather than mathematics, only limited technical evaluation may take place. The analysis of engineering documents uses mathematics to verify that the design is correct. Therefore, if software engineers are to perform as *engineers* then the software documents should be similar to engineering documents, and should include sufficient mathematics to allow rigorous analysis to be performed. The mathematics for software engineering is described in the next section.

1.2 Software Engineering Mathematics

The use of mathematics plays a key role in the engineer's work; for example, bridge designers will develop a mathematical model of a bridge prior to its construction. The model is a simplification of the reality, and an exploration of the model enables a deeper understanding of the proposed system to be gained. Engineers will model the various stresses on the bridge to ensure that the bridge design can deal with the projected traffic flow. The ability to use mathematics to solve practical problems is part of the engineer's education, and part of the daily work of an engineer. The engineer applies mathematics and models to the design of the product, and the analysis of the design is a mathematical activity.

Mathematics allows a rigorous analysis to take place and avoids an overreliance on intuition. The focus needs to be on mathematics that can be applied to solve practical problems and in developing products that are fit for use, rather than in mathematics for its own sake that is the interest to the pure mathematician. The emphasis in engineering is always in the application of the theorem rather than in the proof, and the objective is therefore to teach students how to use and apply mathematics to program well and to solve practical problems.

There is a rich body of classical mathematics available that may be applied to software engineering. Other specialized mathematical methods and notations have been developed by others to assist in software engineering (e.g., Z, VDM, VDM$^{\clubsuit}$, and CSP). The mathematical foundation for software engineering should include:

Area	Description
Set Theory	This material is elementary but fundamental. It is familiar to all high-school students. It includes, for example, set union and intersection operations; the Cartesian product of two sets, etc.
Relations	A relation between A and B is a subset of $A \times B$. For example, the relation T(A, A) where (a_1, a_2) $\in T$ if a_1 is taller than a_2

Functions	A function $f: A \rightarrow B$ is a relation where for each $a \in A$ there is exactly one $b \in B$ such that $(a,b) \in f$. This is denoted by $f(a) = b$. Functions may be total or partial.
Logic	Logic is the foundation for formal reasoning. It includes the study of propositional calculus and predicate calculus. It enables further truths to be derived from existing truths.
Calculus	Calculus is used extensively in engineering and physics to solve practical problems. It includes differentiation and integration, numerical methods, solving differential equations, etc.
Probability Theory	Probability theory is concerned with determining the mathematical probability of various events occurring. One example of its use in software engineering is in predicting the reliability of a software product.
Finite State Machines	Finite state machines are mathematical entities that are employed to model the execution of a program. The mathematical machine is in a given state and depending on the input there is a change to a new state. Finite state machines may be deterministic or non-deterministic.
Tabular Expressions	This is an approach developed by Parnas and others and may be employed to specify the requirements of a system. It allows complex predicate calculus expressions to be presented in a more readable form (in a 2-dimensional table) using a divide and conquer technique.
Graph Theory	Graphs are useful in modeling networks and a graph consists of vertices and edges. An edge joins two vertices.
Matrix Theory	This includes the study of 2 x 2 and m x n dimensional matrices. It includes calculating the determinants of a matrix and the inverses of a matrix.

Table 1.2. Mathematics for Software Engineering

The use of mathematics in computer science is discussed in more detail in Chapter 2. The emphasis is on mathematics that can be applied to solve practical problems rather than in theoretical mathematics. Next, we consider various formal methods that may be employed to assist in the development of high-quality software. Some of these are described in more detail in later chapters.

1.3 Formal Methods

Spivey (cf., chapter one of [Spi:92]) describes "formal specifications" as the use of mathematical notation to describe in a precise way the properties which an information system must have, without unduly constraining the way in which these properties are achieved. Formal methods consist of formal specification languages or notations and generally employ a collection of tools to support the syntax checking of the specification and proof of properties of the specification. This abstraction away from implementation enables questions about what the system does to be answered, independently of the implementation, i.e., the detailed code. Furthermore, the unambiguous nature of mathematical notation avoids the problem of speculation about the meaning of phrases in an imprecisely worded natural language description of a system. Natural language is inherently ambiguous and subject to these limitations, whereas mathematics employs highly precise notation with sound rules of mathematical inference.

The formal specification thus becomes the key reference point for the different parties concerned with the construction of the system. This includes determining customer needs, program implementation, testing of results, and program documentation. It follows that the formal specification is a valuable means of promoting a common understanding for all those concerned with the system. The term *formal methods* is used to describe a formal specification language and a method for the design and implementation of computer systems.

The specification is written in a mathematical language, and the implementation is derived from the specification via step-wise refinement.[11] The refinement step makes the specification more concrete and closer to the actual implementation. There is an associated proof obligation that the refinement is valid, and that the concrete state preserves the properties of the more abstract state. Thus, assuming that the original specification is correct and the proofs of correctness of each refinement step are valid, then there is a very high degree of confidence in the correctness of the implemented software. Step-wise refinement is illustrated as follows: the initial specification S is the initial model M_0; it is then refined into the more concrete model M_1, and M_1 is then refined into M_2, and so on until the eventual implementation $M_n = E$ is produced.

$$S = M_0 \sqsubseteq M_1 \sqsubseteq M_2 \sqsubseteq M_3 \sqsubseteq \ldots\ldots \sqsubseteq M_n = E$$

Requirements are the foundation stone from which the system is built, and irrespective of the best design and development practices, the product will be incorrect and not what the customer wants if the requirements are incorrect. The objective of requirements validation is to ensure that the requirements are correct and reflect what is actually required by the customer (in order to build the right system). Formal methods may be employed to model the requirements, and the model exploration yields further desirable or undesirable properties. The

---

[11] It is questionable whether step-wise refinement is cost effective in mainstream software engineering.

ability to prove that certain properties are true of the specification is very valuable, especially in safety critical and security critical applications. These properties are logical consequences of the definition of the requirements, and, if appropriate, the requirements may need to be amended appropriately. Thus, formal methods may be employed for requirements validation and in a sense to debug the requirements.

The use of formal methods generally leads to more robust software and increased confidence in the correctness of the software. The challenges involved in the deployment of formal methods in an organization include the education of staff in formal specification, as formal specification and the use of mathematical techniques may be a culture shock to many staff.

Formal methods have been applied to a diverse range of applications, including circuit design, artificial intelligence, security critical field, the safety critical field, railways, microprocessor verification, the specification of standards, and the specification and verification of programs.

Formal methods have been criticized by Parnas and others on the following grounds:

No.	Criticism
1.	Often the formal specification (written for example in Z or VDM) is as difficult to read as the program and therefore does not add value.[12]
2.	Many formal specifications are wrong.[13]
3.	Formal methods are strong on syntax but provide little assistance in deciding on what technical information should be recorded using the syntax.[14]
4.	Formal specifications provide a model of the proposed system. However, a precise unambiguous mathematical statement of the requirements is what is needed. It should be clear what is required and what is not required.[15]

---

[12] Of course, others might reply by saying that some of Parnas's tables are not exactly intuitive, and that the notation used in some of the tables is quite unfriendly. The usability of all of the mathematical approaches needs to be enhanced if they are to be taken seriously by industrialists.

[13] Obviously, the formal specification must be analyzed using mathematical reasoning and tools to ensure its correctness. The validation of a formal specification can be carried out using mathematical proof of key properties of the specification; software inspections; or specification animation.

[14] Approaches such as VDM include a method for software development as well as the specification language.

[15] Models are extremely valuable as they allow simplification of the reality. A mathematical study of the model demonstrates whether it is a suitable representation of the system. Models allow properties of the proposed requirements to be studied prior to implementation.

| 5. | Step-wise refinement is unrealistic.[16] It is like, for example, deriving a bridge from the description of a river and the expected traffic on the bridge. Refining a formal specification to a program is like refining a blueprint until it turned into a house. This is hardly realistic and there is always a need for a creative step in design. |
| 6. | Many unnecessary mathematical formalisms have been developed in formal methods rather than using the available classical mathematics.[17] |

Table 1.3. Criticisms of Formal Methods

However, formal methods are potentially quite useful in software engineering. The use of a method such as Z or VDM forces the engineer to be precise and helps to avoid ambiguities present in natural language. My experience is that formal specifications such as Z or VDM are reasonably easy to use. Clearly, a formal specification should be subject to peer review to ensure that it is correct. New formalisms may potentially add value in expressive power but need to be intuitive to be usable by practitioners. The advantage of classical mathematics is that it is familiar to students.

1.3.1 Why Should We Use Formal Methods?

There is a very strong motivation to use best practices in software engineering in order to produce software adhering to high quality standards. Flaws in software may at best cause minor irritations to customers, and in a worst-case scenario could potentially cause major damage to a customer's business or loss of life. Consequently, companies will employ best practices to mature their software processes. Formal methods are one leading-edge technology which studies suggest may be of benefit to companies who wish to minimize the occurrence of defects in software products.

The use of formal methods is mandatory in certain circumstances. The Ministry of Defence in the United Kingdom issued two safety-critical standards in the early 1990s related to the use of formal methods in the software development lifecycle. The first is Defence Standard 0055, i.e., Def Stan 00-55, *The*

---

[16] Step-wise refinement involves rewriting a program ad nauseum (each refinement step produces a more concrete specification that includes code and formal specification) until eventually the detailed code is produced. However, tool support may make refinement easier. The refinement calculus offers a very rigorous approach to develop correct programs from formal specifications. However, it is debatable whether it is cost effective in mainstream software development.

[17] My preference is for the use of classical mathematics for specification. However, I see approaches such as VDM or Z as useful as they do add greater rigor to the software development process than is achieved with the use of natural language. In fact, both Z and VDM are reasonably easy to learn, and there have been some good results obtained by their use. I prefer to avoid fundamentalism that I have seen elsewhere and am of the view that if other notations add value in formal specification then it is perfectly reasonable to employ them.

Procurement of safety critical software in defense equipment [MOD:91a]. This standard makes it mandatory to employ formal methods in safety-critical software development in the UK; and, in particular, it mandates the use of formal mathematical proof that the most crucial programs correctly implement their specifications. The other Defence Standard is Def Stan 00-56 *Hazard analysis and safety classification of the computer and programmable electronic system elements of defense equipment* [MOD:91b]. The objective of this standard is to provide guidance to identify which systems or parts of systems being developed are safety-critical and thereby require the use of formal methods. This is achieved by subjecting a proposed system to an initial hazard analysis to determine whether there are safety-critical parts. The reaction to these defence standards 00-55 and 00-56[18] was quite hostile initially as most suppliers were unlikely to meet the technical and organization requirements of the standard, and this is described in [Tie:91]. The standard indicates how seriously the Ministry of Defence in the United Kingdom takes safety, and Brown in [Bro:90] argues that

Missile systems must be presumed dangerous until shown to be safe, and that the absence of evidence for the existence of dangerous errors does not amount to evidence for the absence of danger.

It is quite possible that a software company may be sued for software which injures a third party, and it is conjectured in [Mac:93] that the day is not far off when

A system failure traced to a software fault and injurious to a third party, will lead to a successful litigation against the developers of the said system software.

This suggests that companies will need a quality assurance program that will demonstrate that every reasonable practice was considered to prevent the occurrence of defects. One such practice for defect prevention is the use of formal methods in the software development lifecycle, and in some domains, e.g., the safety critical domain, it looks likely that the exclusion of formal methods in the software development cycle may need to be justified.

There is evidence to suggest that the use of formal methods provides savings in the cost of the project, for example, an independent audit of the large CICS transaction processing project at IBM demonstrated a 9% cost saving attributed to the use of formal methods. An independent audit of the Inmos floating point unit of the T800 project confirmed that the use of formal methods led to an estimated 12-month reduction in testing time. These savings are discussed in more detail in chapter one of [HB:95].

There is evidence to suggest that the use of formal methods provides savings in the cost of the project, for example, an independent audit of the large CICS transaction processing project at IBM demonstrated a 9% cost saving attributed to the use of formal methods. An independent audit of the Inmos floating point unit of the T800 project confirmed that the use of formal methods led to an estimated 12-month reduction in testing time. These savings are discussed in more detail in chapter one of [HB:95].

The approach of modern software companies to providing high-quality software on time and within budget is to employ a mature software development

---

[18] I understand that the UK Defence Standards 0055 and 0056 have been revised recently to be less prescriptive on the use of formal methods.

process including inspections and testing. Models such as the CMM [Pau:93] and CMMI [CKS:03] are employed to assist the organization to mature its software process. The process-based approach is also useful in that it demonstrates that reasonable practices are employed to identify and prevent the occurrence of defects, and an ISO 9000 certified software development organization has been independently assessed to have reasonable software development practices in place. A formal methods approach is complementary to these models, and for example, it fits comfortably into the defect prevention key process area and the technology change management key process area on the Capability Maturity Model.

1.3.2 Applications of Formal Methods

Formal methods are used in academia and to varying degrees in industry. The safety-critical and security critical fields are two key areas to which formal methods has been successfully applied in industry. Several organizations have piloted formal methods with varying degrees of success. These include IBM, which actually developed VDM at its laboratory in Vienna. Another IBM site, IBM (Hursley) piloted the *Z* formal specification language in the United Kingdom, and it was employed for the CICS (Customer Information Control System) project. This is an online transaction processing system with over 500,000 lines of code. The project generated valuable feedback to the formal methods community, and although it was very successful in the sense that an independent audit verified that the use of formal methods generated a 9% cost saving, there was resistance to the deployment of the formal methods in the organization.[19] This was attributed to the lack of education on formal methods in computer science curricula, lack of adequate support tools for formal methods, and the difficulty that the programmers had with mathematics and logic.

The mathematical techniques developed by Parnas (i.e., requirements model and tabular expressions) have been employed to specify the requirements of the A-7 aircraft as part of a software cost reduction program for the United States Navy.[20] Tabular expressions have also been employed for the software inspection of the automated shutdown software of the Darlington Nuclear power plant in Canada.[21] These are two successful uses of mathematical techniques in software engineering.

---

[19] I recall a keynote presentation by Peter Lupton of IBM (Hursley) at the Formal Methods Europe (FME'93) conference in which he noted that there was a resistance to the use of formal methods among the software engineers at IBM (Hursley), and that the engineers found the Z notation to be slightly unintuitive.

[20] However, I was disappointed to learn that the resulting software was actually never deployed on the A-7 aircraft.

[21] This was an impressive use of mathematical techniques and it has been acknowledged that formal methods must play an important role in future developments at Darlington. However, given the time and cost involved in the software inspection of the shutdown software some managers have less enthusiasm in shifting from hardware to software controllers [Ger:94].

Formal methods have been successfully applied to the hardware verification field; for example, parts of the Viper microprocessor[22] were formally verified, and the FM9001 microprocessor was formally verified by the Boyer Moore theorem prover [HB:95]. There are many examples of the use of formal methods in the railway domain, and examples dealing with the modeling and verification of a railroad gate controller and railway signaling are described in [HB:95]. Clearly, it is essential to verify safety critical properties such as *when the train goes through the level crossing then the gate is closed*. The mandatory use of formal methods in some safety and security-critical fields has led to formal methods being employed to verify correctness in the nuclear power industry, in the aerospace industry, in the security technology area, and the railroad domain. These sectors are subject to stringent regulatory controls to ensure safety and security.

Formal methods have been successfully applied to the telecommunications domain, and have been useful in investigating the feature interaction problem as described in [Bou:94]. The EC SCORE project considered mathematical techniques to identify and eliminate feature interaction in the telecommunications environment. The feature interaction problem occurs when two features that work correctly in isolation fail to work correctly together.

Formal methods have been applied to domains which have little to do with computer science, for example, to the problem of the formal specification of the single transferable voting system in [Pop:97], and to various organizations and structures in [ORg:97]. There is an extensive collection of examples to which formal methods have been applied, and a selection of these are described in detail in [HB:95]. Formal methods have also been applied to the problem of reuse.

1.3.3 Tools for Formal Methods

One of the main criticisms of formal methods is the lack of available or usable tools to support the engineer in writing the formal specification or in doing the proof. Many of the early tools were criticized as being of academic use only and not being of industrial strength, but in recent years better tools have become available to support the formal methods community. The expectation is that more and more enhanced tools will continue to become available to support the engineer's work in formal specification and formal proof.

There are various kinds of tools employed to support the formal software development environment, including syntax checkers to check that the specification is syntactically correct or specialized editors to ensure that the

---

[22] The VIPER microprocessor chip has been very controversial. It is an example of where formal methods were oversold in that the developers RSRE (Royal Signals and Radar Establishment) of the UK and Charter (a company licensed to exploit the VIPER results) claimed that the VIPER chip is formally specified and the chip design conforms to the specification. However, the report by Avra Cohen of Cambridge University's computer laboratories argued that the claim by RSRE and Charter was misleading. Computational Logic of the United States later confirmed Avra Cohn's conclusions.

written specification is syntactically correct; tools to support refinement; automated code generators to generate a high-level language corresponding to the specification; theorem provers to demonstrate the presence or absence of key properties and to prove the correctness of refinement steps, and to identify and resolve proof obligations; and specification animation tools where the execution of the specification can be simulated. Such tools are available from vendors like *B*-Core and IFAD.

The tools are developed to support existing methods, and there is a recent trend toward an integrated set of method and tools, rather than loosely coupled tools; for example, the *B*-Toolkit from *B*-Core is an integrated set of tools that supports the *B*-Method. These include syntax and type checking, specification animation, proof obligation generator, an auto-prover, a proof assistor, and code generation. Thus, in theory, a complete formal development from initial specification to final implementation may be achieved, with every proof obligation justified, leading to a provably correct program.

The IFAD Toolbox[23] is a well-known support tool for the VDM-SL specification language, and it includes support for syntax and type checking, an interpreter and debugger to execute and debug the specification, and a code generator to convert from VDM-SL to C++. It also includes support for graphical notations such as the OMT/UML design notations.

SDL is a specification language which is employed in event driven real time systems. It is an object-orientated graphical formal language, and support for SDL is provided by the SDT tool from Telelogic. The SDT tool provides code generation from the specification into the C or C++ programming languages, and the generated code can be used in simulations as well as in applications. Telelogic provides an ITEX tool which may be used with or independently of SDT. It allows the generation of a test suite from the SDL specification, thereby speeding up the testing cycle.

The RAISE tools are an integrated toolset including the RAISE specification language (RSL) and a collection of tools to support software development including editors and translators from the specification language to Ada or C++. There are many other tools available, including the Boyer Moore theorem prover, the FDR tool for CSP, the CADiZ tool for the Z specification language, the Mural tool for VDM, the LOTOS toolbox for LOTOS specifications, and the PVS tool.

Finally, various research groups are investigating methods and tools to assist the delivery of high-quality software. This includes a group led by Parnas at SQRL, University of Limerick, Ireland[24] that is developing a table tool system to support tabular expressions and to thereby support engineers in specifying requirements. These include tools for the creation of tables; tools to check

---

[23] The IFAD Toolbox has been renamed to VDMTools as IFAD sold the VDM Tools to CSK in Japan. The tools are expected to be available worldwide and will be improved further.

[24] This group is being supported by Science Foundation Ireland with €5-6 million of Irish taxpayers' funds.

consistency and completeness properties of tables; tools to perform table composition and tools to generate a test oracle from a table.

Formal Methods and Reuse

Effective software reuse helps to speed up software development productivity, and this is of particular importance in rapid application software development. The idea is to develop building blocks which may then be reused in future projects, and this requires that the component be of high quality and reliability, that the domains to which the component may be effectively applied be well known, and that a documented description exists of the actual behavior of the component and the circumstances in which it may be employed.[25]

Effective reuse is typically limited to a particular domain, and there are reuse models to assist organizations that may be employed to assess or diagnose the current reuse practices in the organization, as this enables a reuse strategy to be developed and implemented. Systematic reuse is being researched in academia and industry, and the ROADS project was an EC funded project which included the European Software Institute (ESI) and Thompson as partners to investigate a reuse approach for domain-based software. The software product line approach [ClN:02] proposed by the SEI is growing in popularity.

Formal methods have a role to play in software reuse also, as they offer enhanced confidence in the correctness of the component and provide an unambiguous formal description of the behavior of the particular component. The component may be tested extensively to provide extra confidence in its correctness. A component is generally used in many different environments, and the fact that a component has worked successfully in one situation is no guarantee that it will work successfully in the future, as there could be potential undesirable interaction between it and other components, or other software. Consequently, it is desirable that the behavior of the component be unambiguously specified and fully understood, and that a formal analysis of component composition be performed to ensure that risks are minimized and that the resulting software is of a high quality.

There has been research into the formalization of components in both academia and industry. The EC funded SCORE research project conducted as part of the European RACE II program considered the challenge of reuse. It included the formal specification of components and developed a component model. Formal methods have a role to play in identifying and eliminating undesirable component interaction.

---

[25] Parnas has noted that lots of reusable software is developed that nobody reuses. However, this is a key challenge that companies have to face if they wish to reduce their development costs and have faster software development. Reducing development costs and faster delivery are two key drivers in today's competitive environment.

1.3.4 Model-Oriented Approach

There are two key approaches to formal methods: namely the model-oriented approach of VDM or Z, and the algebraic, or axiomatic approach, which includes the process calculi such as the calculus communicating systems (CCS) or communicating sequential processes (CSP).

A model-oriented approach to specification is based on mathematical models. A mathematical model is a mathematical representation or abstraction of a physical entity or system. The representation or model aims to provide a mathematical explanation of the behavior of the system or the physical world. A model is considered suitable if its properties closely match the properties of the system, and if its calculations match and simplify calculations in the real system, and if predictions of future behavior may be made. The physical world is dominated by models, e.g., models of the weather system, that enable predictions of the weather to be made, and economic models that enable predictions of the future performance of the economy may be made.

It is fundamental to explore the model and to consider the behavior of the model and the behavior of the real world entity. The extent to which the model explains the underlying physical behavior and allows predictions of future behavior to be made will determine its acceptability as a representation of the physical world. Models that are ineffective at explaining the physical world are replaced with new models which offer a better explanation of the manifested physical behavior. There are many examples in science of the replacement of one theory by a newer one: the replacement of the Ptolemaic model of the universe by the Copernican model or the replacement of Newtonian physics by Einstein's theories on relativity. The revolutions that take place in science are described in detail in Kuhn's famous work on scientific revolutions [Kuh:70].

A model is a foundation stone from which the theory is built, and from which explanations and justification of behavior are made. It is not envisaged that we should justify the model itself, and if the model explains the known behavior of the system, it is thus deemed adequate and suitable. Thus the model may be viewed as the starting point of the system. Conversely, if inadequacies are identified with the model we may view the theory and its foundations as collapsing, in a similar manner to a house of cards; alternately, we may search for amendments to the theory to address the inadequacies.

The model-oriented approach to software development involves defining an abstract model of the proposed software system. The model acts as a representation of the proposed system, and the model is then explored to assess its suitability. The exploration of the model takes the form of model interrogation, i.e., asking questions and determining the effectiveness of the model in answering the questions. The modeling in formal methods is typically performed via elementary discrete mathematics, including set theory, sequences, and functions.

The modeling approach is adopted by the Vienna Development Method (VDM) and Z. VDM arose from work done in the IBM laboratory in Vienna in formalizing the semantics for the PL/1 compiler, and was later applied to the specification of software systems. The *Z* specification language had its origins in the early 1980s at Oxford University.

VDM is a method for software development and includes a specification language originally named Meta IV (a pun on *metaphor*), and later renamed VDM-SL in the standardization of VDM. The approach to software development is via step-wise refinement. There are several schools of VDM, including VDM^{++}, the object-oriented extension to VDM, and what has become known as the Irish school of VDM, i.e., VDM$^\clubsuit$, which was developed at Trinity College, Dublin.

1.3.5 Axiomatic Approach

The axiomatic approach focuses on the properties that the proposed system is to satisfy, and there is no intention to produce an abstract model of the system. The required properties and underlying behavior of the system are stated in mathematical notation. The difference between the axiomatic specification and a model-based approach is illustrated by the example of a stack. The stack is a well-known structure in computer science, and includes stack operators for pushing an element onto the stack and popping an element from the stack. The properties of *pop* and *push* are explicitly defined in the axiomatic approach, whereas in the model-oriented approach, an explicit model of the stack and its operations are constructed in terms of the effect the operations have on the model. The specification of an abstract data type of a stack involves the specification of the properties of the abstract data type, but the abstract data type is not explicitly defined; i.e., only the properties are defined. The specification of the *pop* operation on a stack is given by axiomatic properties, for example, *pop(push(s,x)) = s.*

The *property-oriented approach* has the advantage that the implementer is not constrained to a particular choice of implementation, and the only constraint is that the implementation must satisfy the stipulated properties. The emphasis is on the identification and expression of the required properties of the system and the actual representation or implementation issues are avoided, and the focus is on the specification of the underlying behavior. Properties are typically stated using mathematical logic or higher-order logics, and mechanized theorem-proving techniques may be employed to prove results.

One potential problem with the axiomatic approach is that the properties specified may not be satisfiable in any implementation. Thus, whenever a "formal theory" is developed a corresponding "model" of the theory must be identified, in order to ensure that the properties may be realized in practice. That is, when proposing a system that is to satisfy some set of properties, there is a need to prove that there is at least one system that will satisfy the set of properties. The model-oriented approach has an explicit model to start with and so this problem does not arise. A constructive approach is preferred by some groups in formal methods, and in this approach whenever existence is stipulated *constructive existence* is implied, where a direct example of the existence of an object can be exhibited, or an algorithm to produce the object within a finite time period exists. This is different from an existence proof, where it is known that there is a solution to a particular problem but there is no algorithm to construct the solution.

1.3.6 The Vienna Development Method

VDM dates from work done by the IBM research laboratory in Vienna. Their aim was to specify the semantics of the PL/1 programming language. This was achieved by employing the Vienna Definition Language (VDL), taking an operational semantic approach; i.e., the semantics of a language is determined in terms of a hypothetical machine which interprets the programs of that language [BjJ:78, BjJ:82]. Later work led to the Vienna Development Method (VDM) with its specification language, Meta IV. This concerned itself with the denotational semantics of programming languages; i.e., a mathematical object (set, function, etc.) is associated with each phrase of the language [BjJ:82]. The mathematical object is the *denotation* of the phrase.

VDM is a *model-oriented approach* and this means that an explicit model of the state of an abstract machine is given, and operations are defined in terms of this state. Operations may act on the system state, taking inputs, and producing outputs and a new system state. Operations are defined in a precondition and postcondition style. Each operation has an associated proof obligation to ensure that if the precondition is true, then the operation preserves the system invariant. The initial state itself is, of course, required to satisfy the system invariant. VDM uses keywords to distinguish different parts of the specification, e.g., preconditions, postconditions, as introduced by the keywords *pre* and *post* respectively. In keeping with the philosophy that formal methods specifies *what* a system does as distinct from *how*, VDM employs postconditions to stipulate the effect of the operation on the state. The previous state is then distinguished by employing *hooked variables*, e.g., $v^{\curvearrowleft}$, and the postcondition specifies the new state (defined by a logical predicate relating the prestate to the poststate) from the previous state.

VDM is more than its specification language Meta IV (called VDM-SL in the standardization of VDM) and is, in fact, a software development method, with rules to verify the steps of development. The rules enable the executable specification, i.e., the detailed code, to be obtained from the initial specification via refinement steps. Thus, we have a sequence $S = S_0, S_1, \ldots, S_n = E$ of specifications, where S is the initial specification, and E is the final (executable) specification. Retrieval functions enable a return from a more concrete specification, to the more abstract specification. The initial specification consists of an initial state, a system state, and a set of operations. The system state is a particular domain, where a domain is built out of primitive domains such as the set of natural numbers, etc., or constructed from primitive domains using domain constructors such as Cartesian product, disjoint union, etc. A domain-invariant predicate may further constrain the domain, and a *type* in VDM reflects a domain obtained in this way. Thus, a type in VDM is more specific than the signature of the type, and thus represents values in the domain defined by the signature, which satisfy the domain invariant. In view of this approach to types, it is clear that VDM types may not be "statically type checked".

VDM specifications are structured into modules, with a module containing the module name, parameters, types, operations, etc. Partial functions occur frequently in computer science as many functions, especially recursively defined functions can be undefined, or fail to terminate for some arguments in their domain. VDM addresses partial functions by employing nonstandard logical operators, namely the logic of partial functions (LPFs) which can deal with undefined operands. The Boolean expression $T \vee \perp = \perp \vee T = T$; i.e., the truth value of a logical or operation is true if at least one of the logical operands is true, and the undefined term is treated as a don't care value.

VDM is a widely used formal method and has been used in industrial strength projects as well as by the academic community. There is tool support available, for example, the IFAD Toolbox.[26] VDM is described in detail in Chapter 5. There are several variants of VDM, including VDM^{++}, the object-oriented extension of VDM, and the Irish school of the VDM, which is discussed in the next section.

1.3.7 VDM*, the Irish School of VDM

The Irish School of VDM is a variant of standard VDM, and is characterized by [Mac:90] its constructive approach, classical mathematical style, and its terse notation. In particular, this method aims to combine the *what* and *how* of formal methods in that its terse specification style stipulates in concise form *what* the system should do; furthermore, the fact that its specifications are constructive (or functional) means that the *how* is included with the *what*. However, it is important to qualify this by stating that the how as presented by VDM* is not directly executable, as several of its mathematical data types have no corresponding structure in high-level programming languages or functional languages. Thus, a conversion or reification of the specification into a functional or higher-level language must take place to ensure a successful execution. It should be noted that the fact that a specification is constructive is no guarantee that it is a good implementation strategy, if the construction itself is naive. This issue is considered (cf. pp. 135-7 in [Mac:90]) in the construction of the Fibonacci series.

The Irish school follows a similar development methodology as in standard VDM and is a model-oriented approach. The initial specification is presented, with initial state and operations defined. The operations are presented with preconditions; however, no postcondition is necessary as the operation is "functionally" (i.e., explicitly) constructed. Each operation has an associated proof obligation; if the precondition for the operation is true and the operation is performed, then the system invariant remains true after the operation. The proof

---

[26] As discussed earlier the VDM Tools from the IFAD Toolbox are now owned by the CSK Group in Japan.

of invariant preservation normally takes the form of *constructive proofs*. This is especially the case for *existence proofs*, in that the philosophy of the school is to go further than to provide a theoretical proof of existence; rather the aim is to exhibit existence constructively.

The emphasis is on constructive existence and the implication of this is that the school avoids the existential quantifier of predicate calculus. In fact, reliance on logic in proof is kept to a minimum, and emphasis instead is placed on equational reasoning. Special emphasis is placed on studying algebraic structures and their morphisms. Structures with nice algebraic properties are sought, and such a structure includes the monoid, which has closure, associativity, and a unit element. The monoid is a very common structure in computer science, and thus it is appropriate to study and understand it. The concept of isomorphism is powerful, reflecting that two structures are essentially identical, and thus we may choose to work with either, depending on which is more convenient for the task in hand.

The school has been influenced by the work of Polya and Lakatos. The former [Pol:57] advocated a style of problem solving characterized by solving a complex problem by first considering an easier subproblem and considering several examples, which generally leads to a clearer insight into solving the main problem. Lakatos's approach to mathematical discovery (cf. [Lak:76]) is characterized by heuristic methods. A primitive conjecture is proposed and if global counterexamples to the statement of the conjecture are discovered, then the corresponding *hidden lemma* for which this global counterexample is a local counter example is identified and added to the statement of the primitive conjecture. The process repeats, until no more global counterexamples are found. A skeptical view of absolute truth or certainty is inherent in this.

Partial functions are the norm in VDM*, and as in standard VDM, the problem is that recursively defined functions may be undefined, or fail to terminate for several of the arguments in their domain. The logic of partial functions (LPFs) is avoided, and instead care is taken with recursive definitions to ensure termination is achieved for each argument. This is achieved by ensuring that the recursive argument is strictly decreasing in each recursive invocation. The $\perp$ symbol is typically used in the Irish school to represent *undefined or unavailable* or *do not care*. Academic and industrial projects have been conducted using the method of the Irish school, but at this stage tool support is limited.

There are proof obligations to demonstrate that the operations preserve the invariant. Proof obligations require a mathematical proof by hand or a machine-assisted proof to verify that the invariant remains satisfied after the operation. VDM* is described in detail in Chapter 6.

1.3.8 The *Z* Specification Language

Z is a formal specification language founded on Zermelo set theory. It is a model-oriented approach with an explicit model of the state of an abstract machine given, and operations are defined in terms of this state. Its main features include a mathematical notation which is similar to VDM and the schema calculus. The latter is visually striking and consists essentially of boxes, with these boxes or schemas used to describe operations and states. The schema calculus enables schemas to be used as building blocks and combined with other schemas.

The schema calculus is a powerful means of decomposing a specification into smaller pieces or schemas. This decomposition helps to ensure that a Z specification is highly readable, as each individual schema is small in size and self-contained. Exception handling may be addressed by defining schemas for the exception cases, and then combining the exception schema with the original operation schema. Mathematical data types are used to model the data in a system, these data types obey mathematical laws. These laws enable simplification of expressions and are useful with proofs.

Operations are defined in a precondition/postcondition style; however, the precondition is implicitly defined within the operation; i.e., it is not separated out as in standard VDM. Each operation has an associated proof obligation to ensure that if the precondition is true, then the operation preserves the system invariant. The initial state itself is, of course, required to satisfy the system invariant. Postconditions employ a logical predicate which relates the prestate to the poststate, the poststate of a variable being distinguished by priming, e.g., v'. Various conventions are employed within Z specification, for example $v?$ indicates that v is an input variable; $v!$ indicates that v is an output variable. The Ξ Op operation indicates that the operation Op does not affect the state; $\Delta\ Op$ indicates that Op is an operation which affects the state.

Many of the data types employed in Z have no counterpart in standard programming languages. It is therefore important to identify and describe the concrete data structures which ultimately will represent the abstract mathematical structures. As the concrete structures may differ from the abstract, the operations on the abstract data structures may need to be refined to yield operations on the concrete data which yield equivalent results. For simple systems, direct refinement (i.e., one step from abstract specification to implementation) may be possible; in more complex systems, deferred refinement is employed, where a sequence of increasingly concrete specifications are produced to yield the executable specification eventually.

Z has been successfully applied in industry, and one of its well-known successes is the CICS project at IBM Hursley in the United Kingdom. Z is described in detail in Chapter 4.

The *B*-Method

The *B-Technologies* (cf. [McD:94]) consist of three components; a method for software development, namely the *B*-Method; a supporting set of tools, namely, the *B*-Toolkit; and a generic program for symbol manipulation, namely, the *B*-Tool (from which the *B*-Toolkit is derived). The *B*-Method is a model-oriented approach and is closely related to the Z specification language. Every construct in the method has a set theoretic counterpart, and the method is founded on Zermelo set theory. Each operation has an explicit precondition, and an immediate proof obligation is that the precondition is stronger than the weakest precondition for the operation.

One key purpose [McD:94] of the *abstract machine* in the B-Method is to provide encapsulation of variables representing the state of the machine and operations which manipulate the state. Machines may refer to other machines, and a machine may be introduced as a refinement of another machine. The abstract machine are specification machines, refinement machines, or implementable machines. The B-Method adopts a layered approach to design where the design is gradually made more concrete by a sequence of design layers, where each design layer is a refinement that involves a more detailed implementation in terms of abstract machines of the previous layer. The design refinement ends when the final layer is implemented purely in terms of library machines. Any refinement of a machine by another has associated proof obligations and proof may be carried out to verify the validity of the refinement step.

Specification animation of the Abstract Machine Notation (AMN) specification is possible with the B-Toolkit and this enables typical usage scenarios of the AMN specification to be explored for requirements validation. This is, in effect, an early form of testing and may be used to demonstrate the presence or absence of desirable or undesirable behavior. Verification takes the form of proof to demonstrate that the invariant is preserved when the operation is executed within its precondition, and this is performed on the AMN specification with the B-Toolkit.

The B-Toolkit provides several tools which support the B-Method, and these include syntax and type checking; specification animation, proof obligation generator, auto prover, proof assistor, and code generation. Thus, in theory, a complete formal development from initial specification to final implementation may be achieved, with every proof obligation justified, leading to a provably correct program.

The B-Method and toolkit have been successfully applied in industrial applications and one of the projects to which they have been applied is the CICS project at IBM Hursley in the United Kingdom. The B-Method and Toolkit have been designed to support the complete software development process from specification to code. The application of B to the CICS project is described in [Hoa:95], and the automated support provided has been cited as a major benefit of the application of the B-Method and the B-Toolkit.

1.3.9 Propositional and Predicate Calculus

Propositional calculus associates a truth-value with each proposition and is widely employed in mathematics and logic. There are a rich set of connectives employed in the calculus for truth functional operations, and these include $A \Rightarrow B$, $A \wedge B$, $A \vee B$ which denote, respectively, the conditional if A then B, the conjunction of A and B, and the disjunction of A and B. A truth table may be constructed, and the truth values are normally the binary values of *true* and *false.* There are other logics, for example, the logic of partial functions that is a 3-valued logic. These allow an undefined truth-value for the proposition.

Predicate calculus includes variables and a formula in predicate calculus is built up from the basic symbols of the language; these symbols include variables; predicate symbols, including equality; function symbols, including the constants; logical symbols, e.g., ∃, ∧, ∨, ¬, etc.; and the punctuation symbols, e.g., brackets and commas. The formulae of predicate calculus are built from terms, where a *term* is a key construct, and is defined recursively as a variable or individual constant or as some function containing terms as arguments. A formula may be an atomic formula or built from other formulae via the logical symbols. Other logical symbols are then defined as abbreviations of the basic logical symbols.

An interpretation gives meaning to a formula. If the formula is a sentence (i.e., does not contain any free variables) then the given interpretation is true or false. If a formula has free variables, then the truth or falsity of the formula depends on the values given to the free variables. A free formula essentially describes a relation say, $R(x_1, x_2, \ldots x_n)$ such that $R(x_1, x_2, \ldots x_n)$ is true if $(x_1, x_2, \ldots x_n)$ is in relation R. If a free formula is true irrespective of the values given to the free variables, then the formula is true in the interpretation.

A valuation (meaning) function is associated with the interpretation, and gives meaning to the connectives. Thus, associated with each constant c is a constant c_Σ in some universe of values Σ; with each function symbol f, we have a function symbol f_Σ in Σ; and for each predicate symbol P a relation P_Σ in Σ. The valuation function in effect gives a semantics to the language of the predicate calculus L. The truth of a proposition P is then defined in the natural way, in terms of the meanings of the terms, the meanings of the functions, predicate symbols, and the normal meanings of the connectives.

Mendelson (cf. p. 48 of [Men:87]) provides a rigorous though technical definition of truth in terms of satisfaction (with respect to an interpretation M). Intuitively a formula F is *satisfiable* if it is *true* (in the intuitive sense) for some assignment of the free variables in the formula F. If a formula F is satisfied for every possible assignment to the free variables in F, then it is *true* (in the technical sense) for the interpretation M. An analogous definition is provided for *false* in the interpretation M.

A formula is *valid* if it is true in every interpretation; however, as there may be an uncountable number of interpretations, it may not be possible to check this requirement in practice. M is said to be a model for a set of formulae if and only if every formula is true in M.

There is a distinction between proof theoretic and model theoretic approaches in predicate calculus. *Proof theoretic* is essentially syntactic, and we have a list of axioms with rules of inference. In this way the theorems of the calculus may be logically derived and thus we may logically derive (i.e., ⊢ A) the theorems of the calculus. In essence the logical truths are a result of the syntax or form of the formulae, rather than the *meaning* of the formulae. *Model theoretical*, in contrast is essentially semantic. The truths derive essentially from the meaning of the symbols and connectives, rather than the logical structure of the formulae. This is written as ⊨$_M$ A.

A calculus is *sound* if all the logically valid theorems are true in the interpretation, i.e., proof theoretic $\Rightarrow$ model theoretic. A calculus is complete if all the truths in an interpretation are provable in the calculus, i.e., model theoretic $\Rightarrow$ proof theoretic. A calculus is *consistent* if there is no formula A such that $\vdash A$ and $\vdash \neg A$. Logic is discussed in detail in Chapter 3.

Predicate Transformers and Weakest Preconditions

The precondition of a program S is a predicate, i.e., a statement that may be true or false, and it is usually required to prove that if Q is true, where Q is the precondition of a program S; i.e., ($\{Q\}$ S $\{R\}$), then execution of S is guaranteed to terminate in a finite amount of time in a state satisfying R.

The weakest precondition (cf. p. 109 of [Gri:81]) of a command S with respect to a postcondition R represents the set of all states such that if execution begins in any one of these states, then execution will terminate in a finite amount of time in a state with R true. These set of states may be represented by a predicate Q', so that $wp(S,R) = wp_S(R) = Q'$, and so wp_S is a predicate transformer, i.e., it may be regarded as a function on predicates. The weakest precondition is the precondition that places the fewest constraints on the state than all of the other preconditions of (S,R). That is, all of the other preconditions are stronger than the weakest precondition.

The notation $Q\{S\}R$ is used to denote partial correctness and indicates that if execution of S commences in any state satisfying Q, and if execution terminates, then the final state will satisfy R. Often, a predicate Q which is stronger than the weakest precondition $wp(S,R)$ is employed, especially where the calculation of the weakest precondition is nontrivial. Thus, a stronger predicate Q such that $Q \Rightarrow wp(S,R)$ is sometimes employed in these cases.

There are many properties associated with the weakest preconditions and these are used in practice to simplify expressions involving weakest preconditions and in determining the weakest preconditions of various program commands, e.g., assignments, iterations, etc. These are discussed in more detail in Chapter 7. Weakest preconditions are useful in developing a proof of correctness of a program in parallel with its development.

An imperative program may be regarded as a predicate transformer. This is since a predicate P characterizes the set of states in which the predicate P is true, and an imperative program may be regarded as a binary relation on states, which may be extended to a function F, leading to the Hoare triple $P\{F\}Q$. That is, the program F acts as a predicate transformer. The predicate P may be regarded as an input assertion, i.e., a Boolean expression which must be true before the program F is executed. The Boolean expression Q is the output assertion, and is true if the program F terminates, having commenced in a state satisfying P.

The Process Calculi

The objectives of the process calculi [Hor:85] are to provide mathematical models which provide insight into the diverse issues involved in the specification, design, and implementation of computer systems which continuously act and interact with their environment. These systems may be decomposed into subsystems which interact with each other and their environment. The basic building block is the *process*, which is a mathematical abstraction of the interactions between a system and its environment. A process which lasts indefinitely may be specified recursively. Processes may be assembled into systems, execute concurrently, or communicate with each other. Process communication may be synchronized, and generally take the form of a process outputting a message simultaneously to another process inputting a message. Resources may be shared among several processes. Process calculi enrich the understanding of communication and concurrency, and elegant formalisms such as CSP [Hor:85] and CCS [Mil:89] which obey a rich collection of mathematical laws, have been developed.

The expression *(a → P)* in CSP describes a process which first engages in event *a,* and then behaves as process *P*. A recursive definition is written as $(\mu X) \cdot F(X)$ and an example of a simple chocolate vending machine is

$$\text{VMS} = \mu X : \{\text{coin, choc}\} \cdot (\text{coin} \rightarrow (\text{choc} \rightarrow X))$$

The simple vending machine has an alphabet of two symbols, namely, *coin* and *choc*, and the behavior of the machine is that a coin is entered into the machine and then a chocolate selected and provided.

CSP processes use channels to communicate values with their environment, and input on channel *c* is denoted by *(c?.x → P_x)*, which describes a process that accepts any value *x* on channel *c*, and then behaves as process P_x. In contrast, *(c!e → P)* defines a process which outputs the expression *e* on channel *c* and then behaves as process *P*. CSP is discussed in more detail in Chapter 7.

The π-calculus is based on names. Communication between processes takes place between known channels, and the name of a channel may be passed over a channel. There is no distinction between channel names and data values in the π-calculus, and this is a difference between π-calculus and CCS. The output of a value *v* on channel *a* is given by $\bar{a}v$; i.e., output is a negative prefix. Input on a channel *a* is given by *a(x)*, and is a positive prefix. Private links or restrictions are given by *(x)P* in the π-calculus and *P\x* in CCS.

1.3.10 The Parnas Way

David L. Parnas has been influential in the computing field, and his ideas on the specification, design, implementation, maintenance, and documentation of computer software remain important. He advocates a solid engineering approach to the development of high-quality software and argues that the role of engineers is to apply scientific principles and mathematics to design and develop useful products. He argues that computer scientists should be educated as engineers and provided with the right scientific and mathematical background to do their work effectively. His contributions to software engineering include:

- *Tabular Expressions*
Tabular expressions are the mathematical tables for specifying requirements and are also used in design. They enable complex predicate logic expressions to be represented in a simpler form.

- *Mathematical Documentation*
Parnas advocates the use of mathematical documents for software engineering that are precise and complete.

- *Requirements Specification*
He advocates the use of mathematical relations to specify the requirements precisely.

- *Software Design*
His contribution to software design includes information hiding that allows software to be designed for changeability. A module is characterized by its knowledge of a design decision (secret) that it hides from all others. Every information-hiding module has an interface that provides the only means to access the services provided by the modules. The interface hides the module's implementation. Information hiding[27] is used in object-oriented programming.

- *Software Inspections*
His approach to software inspections is quite distinct from the popular approach of the well-known Fagan inspection methodology. The reviewers are required to take an active part in the inspection and they are provided with a list of questions by the author. The reviewers are required to provide documentation of their analysis to justify the answers to the individual questions. The inspections involve the production of mathematical tables, and may be applied to the actual software or documents.

- *Predicate Logic*
Parnas has introduced an approach to deal with undefined values in predicate logic expressions. The approach is quite distinct from the logic of partial functions developed by Cliff Jones.

The Parnas approach to software engineering is discussed in Chapter 8.

---

[27] I see information hiding as the greatest achievements of Parnas. I find it surprising that many in the object-oriented world seem unaware that information hiding goes back to the early 1970s and many have never heard of Parnas.

1.3.11 Unified Modeling Language

The unified modeling language (UML) is a visual modeling language for software systems and facilitates the understanding of the architecture of the system and in managing the complexity of large systems. It was developed by Jim Rumbaugh, Grady Booch, and Ivar Jacobson [Jac:99a] as a notation for modeling object-oriented systems.

UML allows the same information to be presented in many different ways, and there are several UML diagrams providing different viewpoints of the system. Use cases describe scenarios or sequences of actions for the system from the user's viewpoint. A simple example is the operation of an ATM machine. The typical user operations at an ATM machine include the balance inquiry operation, the withdrawal of cash, and the transfer of funds from one account to another. UML includes use case diagrams to express these scenarios.

Class and object diagrams are a part of UML and the object diagram is related to the class diagram in that the object is an instance of the class. There will generally be several objects associated with the class. The class diagram describes the data structure and the allowed operations on the data structure. The concept of class and objects are taken from object-oriented design. Two key classes are customers and accounts for an ATM system, and this includes the data structure for customers and accounts, and also the operations on customers and accounts. The operations include adding or removing a customer and operations to debit or credit an account. The objects of the class are the actual customers of the bank and their corresponding accounts.

Sequence diagrams show the interaction between objects/classes in the system for each use case. The sequences of interactions between objects for an ATM operation to check the balance of an account is illustrated in a sequence diagram that illustrates:

1. Customer inserts the card into the ATM machine.
2. PIN number is requested by the ATM machine.
3. The customer then enters the PIN number.
4. The ATM machine contacts the bank for verification of the number.
5. The bank confirms the validity of the number and the customer then selects the balance inquiry.
6. The ATM contacts the bank to request the balance of the particular account and the bank sends the details to the ATM machine.
7. The balance is displayed on the screen of the ATM machine.
8. The customer then withdraws the card.

UML activity diagrams are similar to flow charts. They are used to show the sequence of activities in a use case and include the specification of decision branches and parallel activities. The sequence of activities for the ATM operation to check the balance of an account may be shown in an activity diagram that illustrates:

1.   Card insertion
2.   Wait for PIN to be entered.
3.   Validate PIN.
4.   If Valid then check balance on account and Display balance.
5.   Otherwise return to 1.

State diagrams (or state charts) show the dynamic behavior of a class and how different operations result in a change of state. There is an initial state and a final state, and the different operations result in different states being entered and exited.

There are several other UML diagrams including the collaboration diagram which is similar to the sequence diagram except that the sequencing is shown via a number system. UML offers a rich notation to model software systems and to understand the proposed system from different viewpoints. The main advantages of UML are:

Advantages of UML
State of the art visual modeling language with a rich expressive notation.
Study of the proposed system before implementation
Visualization of architecture design of the system.
Mechanism to manage complexity of a large system.
Visualization of system from different viewpoints.
Enhanced understanding of implications of user behavior.
Use cases allow description of typical user behavior.
A mechanism to communicate the proposed behavior of the software system. This describes what it will do and what to test against.

Table 1.4. Advantages of UML

UML is discussed in detail in Chapter 10.

Miscellaneous Specification Languages

The RAISE (Rigorous Approach to Industrial software Engineering) project was a European ESPRIT-funded project. Its objective [Geo:91] was to produce a method for the rigorous development of software, based on a wide-spectrum specification language, and accompanied by tool support. It considered standard VDM to be deficient, in that it lacked modularity, and was unable to deal with concurrency. The RAISE specification language (RSL) is designed to address these deficiencies, and an algebraic approach is adopted. Comprehensive support is available from the RAISE tools.

The RAISE method (as distinct from its specification language) covers the software lifecycle, from requirements analysis to code generation. This is achieved via a number of design steps, in which the specification is gradually made more concrete, until ultimately a specification that may be transferred into code is reached. The RAISE toolset includes library tools for storing and retrieving modules and translators from subsets of RSL into Ada and C++.

The Specification and Descriptive Language (SDL) was developed to allow the behavior of telecommunication systems to be described and specified. It may be used at several levels of abstraction, ranging from a very broad overview of a system to detailed design. The behavior of the system is considered as the combined behavior of the processes in the system, and the latter is considered to be an extended finite state machine, i.e., a finite state machine that can use and manipulate data stored in variables local to the machine. Processes may cooperate via signals (i.e., discrete messages) and exhibit deterministic behavior.

A graphical language is employed to describe processes and this involves graphical representation of states, input, output, and decisions. Channels enable communication between blocks (containing processes) and the system (containing blocks connected by channels) and its environment. SDL supports time constraints via the timer construct. The graphical language has a corresponding equivalent textual representation.

SSADM is a structured systems analysis and design method. It presents three distinct views of an information system. These include logical data structures, data flow diagrams, and entity life histories. The behavior of the system is explained by employing a graphical language of symbols; these symbols may indicate a one-to-many relationship, an optional occurrence, mutually exclusive choices, etc. The method is data driven, with emphasis placed on the processes which manipulate the data. User involvement and commitment to the development is emphasized from the earliest stage of the project.

1.3.12 Proof and Formal Methods

The word *proof* has several connotations in various disciplines; for example, in a court of law, the defendant is assumed innocent until proven guilty. The proof of the guilt of the defendant may take the form of certain facts in relation to the movements of the defendant, the defendant's circumstances, the defendant's alibi, statements from witnesses, rebuttal arguments from the defense, and certain theories produced by the prosecution or defense. Ultimately, in the case of a trial by jury, the defendant is judged guilty or not guilty depending on the extent to which the jury has been convinced by the arguments proposed by prosecution and defense.

A mathematical proof typically includes natural language and mathematical symbols; often many of the tedious details of the proof are omitted. The strategy of proof in proving a conjecture tends to be *a divide and conquer* technique, i.e., breaking the conjecture down into subgoals and then attempting to prove the subgoals. Most proofs in formal methods are concerned with cross-checking on the details of the specification or validity of refinement proofs, or proofs that certain properties are satisfied by the specification. There are many tedious lemmas to be proved and theorem provers[28] assist and are essential. Machine proof needs to be explicit and reliance on some brilliant insight is avoided. Proofs by hand are notorious for containing errors or jumps in reasoning, as discussed in chapter one of [HB:95], while machine proofs are extremely lengthy and unreadable, but generally help to avoid errors and jumps in proof as every step needs to be justified.

One well-known theorem prover is the Boyer/Moore theorem prover [BoM:79], and a mathematical proof consists of a sequence of formulae where each element is either an axiom or derived from a previous element in the series by applying a fixed set of mechanical rules. There is an interesting case in the literature concerning the proof of correctness of the VIPER microprocessor[29] [Tie:91] and the actual machine proof consisted of several million formulae.

Theorem provers are invaluable in resolving many of the thousands of proof obligations that arise from a formal specification, and it is not feasible to apply formal methods in an industrial environment without the use of machine assisted proof. Automated theorem proving is difficult, as often mathematicians prove a theorem with an initial intuitive feeling that the theorem is true. Human intervention to provide guidance or intuition improves the effectiveness of the theorem prover.

The proof of various properties about the programs increases confidence in the correctness of the program. However, an absolute proof of correctness is unlikely except for the most trivial of programs. A program may consist of legacy software which is assumed to work, or be created by compilers which are assumed to work; theorem provers are programs which are assumed to function correctly. In order to be absolutely certain one would also need to verify the hardware, customized-off-the-shelf software, subcontractor software, and every single execution path that the software system will be used for. The best that formal methods can claim is increased confidence in correctness of the software.

1.4 Organization of This Book

This chapter provided an introduction to an engineering approach to software quality that places emphasis on the use of mathematics. It included a review of the popular formal methods in the literature. The second chapter considers the mathematical foundation that is required for sound software engineering. The mathematics discussed includes discrete mathematics such as set theory, functions and relations; propositional and predicate logic for software engineers;

---

[28] Most existing theorem provers are difficult to use and are for specialist use only. There is a need to improve the usability of theorem provers.

[29] As discussed earlier this verification was controversial with RSRE and Charter overselling VIPER as a chip design that conforms to its formal specification.

tabular expressions ain software engineering; probability and applied statistics for predicting software reliability; calculus and matrix theory; finite state machines; and graph theory.

Chapter 3 is a detailed examination of mathematical logic including propositional and predicate calculus, as well as considering ways of dealing with undefined values that arise in specification. The next three chapters are concerned with the model-oriented approach of formal specification. The chapter on Z includes the main features of the Z specification language as well as the schema calculus. The chapter on VDM describes the history of its development at the IBM research laboratory in Vienna, as well as the main features of VDM-SL. The chapter on VDM* explains the philosophy of the Irish school of VDM, and explains how it differs from standard VDM. The two most widely used formal specification languages are Z and VDM.

Chapter seven discusses the contribution of Dijkstra and Hoare including the calculus of weakest preconditions developed by Dijkstra, and the axiomatic semantics of programming languages developed by Hoare. Chapter eight discusses the classical engineering approach of Parnas, and includes material on tabular expressions, requirements specification and design, and software inspections.

Chapter 9 examines the Cleanroom approach of Harlan Mills and the mathematics of software reliability. Cleanroom enables a mathematical prediction of the software reliability to be made based on the expected usage of the software. The software reliability is expressed in terms of the mean time to failure (MTTF). Chapter 10 examines the unified modeling language (UML). This is a visual approach to the specification and design of software. The final chapter examines technology transfer of formal methods to an organization.

1.5 Summary

Software engineering involves multiperson construction of multiversion programs. Software engineers need to receive an appropriate engineering education in mathematics and design in order to be able to build high-quality and safe products. Computer science courses tend to include a small amount of mathematics, whereas mathematics is a significant part of an engineering course. The engineering approach to the teaching of mathematics is to emphasize its application and especially the application to developing and analyzing product designs. The mathematics that software engineering students need to be taught includes sets, relations, functions, mathematical logic, tabular expression, and finite state machines. The emphasis is on the application of mathematics to solve practical problems.

Sound software engineering requires the engineer to state precisely the requirements that the software product is to satisfy and then to produce designs that will meet these requirements. Software engineers should start with a precise description of the problem to be solved; then proceed to producing a design and validating the correctness of the design; finally, implementation and testing are performed. An engineering analysis of the design includes mathematics and software inspections, and this is essential to ensure the correctness of the design.

Software engineers have individual responsibilities as professionals. They are responsible for designing and implementing high-quality and reliable software that is safe to use. They are also accountable for their own decisions and actions and have a responsibility to object to decisions that violate professional standards. Professional engineers have a duty to their clients to ensure that they are solving the real problem of the client. They need to precisely state the problem before working on its solution. Engineers need to be honest about current capabilities when asked to work on problems that have no appropriate technical solution rather than accepting a contract for something that cannot be done.

Formal specifications describe in a precise way the requirements of a proposed system. The objective is to specify the program in a mathematical language and to demonstrate that certain properties are satisfied by the specification using mathematical proof. The ultimate objective is to provide confidence that the implementation satisfies the requirements. Formal methods offers increased precision but cannot provide a guarantee of correctness.

The formal methods community have developed a comprehensive collection of methods and tools to assist in the formal specification of software and to prove properties of the software. These include the well-known model based approaches such as VDM and Z, and the axiomatic approaches of Dijkstra and Hoare. The safety-critical field is a domain to which formal methods are quite suited, as they may be applied to verify that stringent safety and reliability properties hold. Tool support is essential for formal methods to be taken seriously by industrialists, and better tools have been provided in recent years by organizations such as *B*-Core and IFAD.

The role of proof in formal methods was discussed, and tool support is essential for industrial proof. The proofs include invariant preservation of operations and the proof of validity of the refinement step. However, the first step in implementing formal methods is to consider formal specification, and the use of mathematical proof and theorem provers belongs to a more advanced deployment of formal methods. The mathematics for software engineering is described in the next chapter.

2
Software Engineering Mathematics

2.1 Introduction

The ability to use mathematics is a differentiator between engineers and technicians. Engineers are taught practical mathematics and apply their mathematical knowledge to solve practical problems, whereas the technician's mathematical education is more limited. The classical engineer applies mathematics and mathematical models to the design of the product, and a classical engineering analysis of the design is a mathematical activity.

The advantage of mathematics is that it allows rigorous analysis and avoids an overreliance on intuition. Mathematics provides precise unambiguous statements, and the mathematical proof of a theorem provides a high degree of confidence in its correctness. The focus in engineering is on mathematics that can be applied to solve practical problems and in developing products that are fit for use, whereas the interest of the pure mathematician is in mathematics for its own sake. The emphasis in engineering is always in the application of the theorem rather than in the proof, and the objective of engineering mathematics is to teach students on using and applying mathematics to program well and to solve practical problems.

Many programming courses treat the mathematics of programming as if it is too difficult or irrelevant to the needs of computer science students. Instead, many programming courses often focus on teaching the latest programming language rather than in placing the emphasis on the design of useful products. Some advocate commencement of programming prior to understanding of problem and this is clearly wrong from a classical engineering viewpoint. It is important to teach students problem-solving skills such as formulating the problem; decomposing a problem into smaller problems; and integrating the solution.

The mathematics discussed in this chapter includes discrete mathematics; set theory; functions; relations; graph theory; calculus; logic; tabular expressions; numerical analysis; probability, and applied statistics.

2.2 Set Theory

A set is a collection of well-defined objects that contains no duplicates. For example, the set of natural numbers $\mathbf{N}$ is an infinite set consisting of the numbers 1, 2, …, and so on. Most sets encountered in computer science are finite as computers can only deal with finite entities. Set theory is a fundamental building block of mathematics and is familiar to all high-school students. Venn diagrams are often employed to give a pictorial representation of a set and the various set operations.

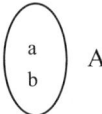

The elements of a finite set S are denoted by $\{x_1, x_2, \ldots x_n\}$. The expression $x \in S$ denotes set membership and indicates that the element x is a member of the set S. A set S is a subset of a set T (denoted $S \subseteq T$) if whenever $s \in S$ then $s \in T$. The set T is a superset of a set S if $S \subseteq T$.

The Cartesian product of sets S and T (denoted $S \times T$) is the set of ordered pairs $\{(s,t) \mid s \in S, t \in T\}$. Clearly, $S \times T \neq T \times S$ and so the Cartesian product is not commutative. The Cartesian product may be extended to n sets $S_1, S_2, \ldots, S_n$ and $S_1 \times S_2 \times .. \times S_n$ is the set of ordered tuples $\{(s_1, s_2, \ldots, s_n) \mid s_1 \in S_1, s_2 \in S_2, .., s_n \in S_n\}$. The Cartesian product allows new sets to be created from existing sets. It is named after the French mathematician Descartes.

A new set may be created from an existing set by the use of a predicate to restrict membership of the set. For example, the even natural numbers are given by $\{x \mid x \in \mathbf{N} \wedge even(x)\}$. In this example, $even(x)$ is a predicate that is true if x is even and false otherwise. In general, $A = \{x \in E \mid P(x)\}$ denotes a set A formed from a set E using the predicate P.

The power set of a set A (denoted $\mathbb{P}A$) denotes the set of subsets of A. There are $2^{|A|}$ elements in the power set of A. For example, the power set of the set A = $\{1,2,3\}$ has 8 elements and is given by:

$$\mathbb{P}A = \{\varnothing, \{1\}, \{2\}, \{3\}, \{1,2\}, \{1,3\}, \{2,3\}, \{1,2,3\}\}.$$

The empty set is denoted by $\varnothing$ and clearly $\varnothing$ is a subset of every set. Two sets A and B are equal if they contain identical elements: i.e., $A = B$ if and only if $A \subseteq B$ and $B \subseteq A$. The singleton set containing just one element x is denoted by $\{x\}$, and clearly $x \in \{x\}$ and $x \neq \{x\}$. Clearly, $y \in \{x\}$ if and only if $x = y$.

The union of two sets A and B is denoted by $A \cup B$. It results in a set that contains all the members of A and of B. It is given by:

$$A \cup B = \{r \mid r \in A \text{ or } r \in B\}.$$

For example, suppose $A = \{1,2,3\}$ and $B = \{2,3,4\}$ then $A \cup B = \{1,2,3,4\}$. Set union is a commutative operation. The intersection of two sets A and B is denoted by $A \cap B$. It results in a set containing the elements that A and B have in common. It is given by:

$$A \cap B = \{r \mid r \in A \text{ and } r \in B\}.$$

Suppose $A = \{1,2,3\}$ and $B = \{2,3,4\}$ then $A \cap B = \{2,3\}$. Set intersection is a commutative operation. The set difference operation $A \setminus B$ yields the elements in A that are not in B. It is given by

$$A \setminus B = \{a \mid a \in A \text{ and } a \notin B\}.$$

For A and B above, $A \setminus B = \{1\}$ and $B \setminus A = \{4\}$. Clearly, set difference is not commutative. The symmetric difference of two sets A and B is denoted by $A \triangle B$ and is given by:

$$A \triangle B = A \setminus B \cup B \setminus A.$$

The symmetric difference operation is commutative. The following illustrates some of the set theoretical operations using Venn diagrams.

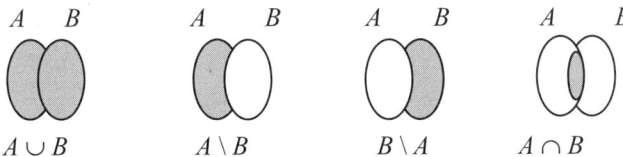

$A \cup B$ $A \setminus B$ $B \setminus A$ $A \cap B$

The complement of a set A (with respect to the universal set U) is given by A^c. This is given by:

$$A^c = \{u \mid u \in U \text{ and } u \notin A\}.$$

The complement of the set A is illustrated by the shaded area below and A^c is simply $U \setminus A$.

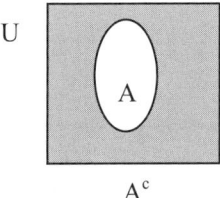

A^c

Next, various properties of set union and set intersection operators are considered. These operators are associative, commutative, and distribute over each other. These properties are listed below:

Property	Description
Commutative	Union and intersection operations are commutative: i.e., $$S \cup T = T \cup S$$ $$S \cap T = T \cap S$$
Associative	Union and intersection operations are associative: i.e., $$R \cup (S \cup T) = (R \cup S) \cup T$$ $$R \cap (S \cap T) = (R \cap S) \cap T$$
Identity	The identity under set union is $\varnothing$ and the identity under intersection is U. $$S \cup \varnothing = S = \varnothing \cup S$$ $$S \cap U = S = U \cap S$$
Distributive	The union operator distributes over the intersection operator and vice versa. E.g., $$R \cap (S \cup T) = (R \cap S) \cup (R \cap T)$$ $$R \cup (S \cap T) = (R \cup S) \cap (R \cup T).$$
De Morgan's Law	The complement of $S \cup T$ is given by: $$(S \cup T)^c = S^c \cap T^c$$ The complement of $S \cap T$ is given by: $$(S \cap T)^c = S^c \cup T^c$$

Table 2.1. Properties of Set Operations

Union and intersection are binary operations but can be extended to generalized union and intersection operations, for example, $\cap_{i=1}^{n} A_i$ denotes the intersection of n sets, and $\cup_{i=1}^{n} A_i$ denotes the union of n sets. De Morgan's law is named after the 19th century English mathematician Augustus De Morgan.[1]

There are many well-known examples of sets including the set of natural numbers that is denoted by **N**; the set of integers is denoted by **Z**; the set of rational numbers is denoted by **Q**; the set of real numbers is denoted by **R**; and the set of complex numbers is denoted by **C**.

---

[1] De Morgan introduced the term mathematical induction and is well-known for his work on mathematical logic. He corresponded with Sir Rowan Hamilton (the Irish mathematician who discovered Quaternions), George Boole (one of the fathers of computing and well known for Boolean algebra, Charles Babbage (one of the fathers of computing and well known for his work on the design and drawings of the analytic engine which was a calculating machine that was to be the forerunner of the modern computer), and Lady Ada Lovelace who in a sense was the first programmer with her ideas that the analytic engine may potentially be employed to produce graphics or solve scientific problems.

2.3 Relations

A binary relation R (A,B), where A and B are sets, is a subset of $A \times B$. The domain of the relation is A and the co-domain of the relation is B. The notation aRb signifies that $(a,b) \in R$. An n-ary relation R $(A_1, A_2, ... A_n)$ is a subset of $(A_1 \times A_2 \times ... \times A_n)$. However, an n-ary relation may also be regarded as a binary relation $R(A,B)$ with $A = A_1 \times A_2 \times ... \times A_{n-1}$ and $B = A_n$. A relation $R(A, A)$ is a relation on A. There are many examples of relations: for example, the relation on a set of people where $(a,b) \in R$ if the height of a is greater than or equal to the height of b. A relation $R(A,B)$ may be expressed pictorially as described below. The arrow from a to p and from a to r indicates that (a,p) and (a,r) are in the relation R. Thus the relation R is given by $\{(a,p), (a,r), (b,q)\}$.

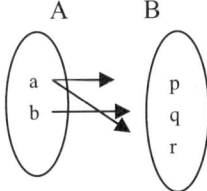

A relation on a set A is reflexive if $(a,a) \in R$ for all $a \in A$. A relation R is symmetric if whenever $(a,b) \in R$ then $(b,a) \in R$. A relation is transitive if whenever $(a,b) \in R$ and $(b,c) \in R$ then $(a,c) \in R$. A relation that is reflexive, symmetric and transitive is termed an equivalence relation.

An equivalence relation on A gives rise to a partition of A where the equivalence classes are given by $\text{Class}(a) = \{x \mid x \in A \text{ and } (a,x) \in R\}$. Similarly, a partition gives rise to an equivalence relation R, where $(a,b) \in R$ if and only if a and b are in the same partition.

The relation below is reflexive, i.e., for each $a \in A$ then $(a,a) \in R$. Symmetric and transitive relations may be illustrated pictorially also.

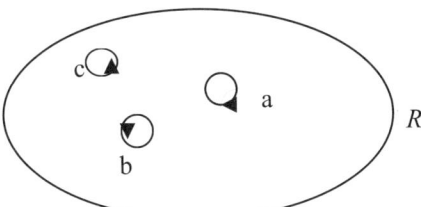

Fig. 2.1. Reflexive Relation

The domain of a relation R (A,B) is given by $\{a \in A \mid \exists b \in B \text{ and } (a,b) \in R\}$. It is denoted by **dom** R. The domain of the relation $R = \{(a,p), (a,r), (b,q)\}$ is $\{a,b\}$. The range of a relation R (A, B) is given by $\{b \in B \mid \exists a \in A \text{ and } (a,b) \in R\}$. It is denoted by **rng** R. The range of the relation $R = \{(a,p), (a,r), (b,q)\}$ is $\{p, q, r\}$.

The composition of two relations $R_1(A,B)$ and $R_2(B,C)$ is given by R_2 o R_1 where $(a,c) \in R_2$ o R_1 if and only there exists $b \in B$ such that $(a,b) \in R_1$ and $(b,c) \in R_2$. The composition of relations is associative: i.e.,

$$(R_3 o\ R_2)\ o\ R_1 = R_3\ o\ (R_2\ o\ R_1)$$

The union of two relations $R_1(A,B)$ and $R_2(A,B)$ is meaningful as these are both subsets of $A \times B$. It is given by $R_1 \cup R_2$ where $(a,b) \in R_1 \cup R_2$ if and only if $(a,b) \in R_1$ or $(a,b) \in R_2$. Similarly, the intersection of R_1 and R_2 is meaningful and is given by $R_1 \cap R_2$ where $(a,b) \in R_1 \cap R_2$ if and only if $(a,b) \in R_1$ and $(a,b) \in R_2$. The relation R_1 is a subset of R_2 ($R_1 \subseteq R_2$) if whenever $(a,b) \in R_1$ then $(a,b) \in R_2$. The inverse of the relation R is given by the relation R^{-1} where:

$$R^{-1} = \{(b,a) \mid (a,b) \in R\}.$$

The composition of R and R^{-1} yields: R^{-1} o $R = \{(a,a) \mid a \in$ dom R$\}$ and R o R^{-1} $= \{(b,b) \mid b \in$ **dom** $R^{-1}\}$. Given a relation R on A then $R^2 = R$ o R is a relation on A. This can be generalized to a relation R^n on A where $R^n = R$ o R o ... o R (n-times). The transitive closure of the relation R on A is given by:

$$R^* = \cup_{i=0}^{\infty} R^i = R^0 \cup R^1 \cup R^2 \cup ...\ R^n \cup ...$$

where R^0 is the reflexive relation containing only each element in the domain of R: i.e., $R^0 = \{(a,a) \mid a \in$ **dom** R$\}$. The positive transitive closure is similar to the transitive closure except that it does not contain R^0. It is given by:

$$R^+ = \cup_{i=1}^{\infty} R^i = R^1 \cup R^2 \cup ...\ R^n \cup ...$$

Parnas has introduced the limited domain relation (LD-relation) where a LD relation L consists of an ordered pair (R_L, C_L) where R_L is a relation and C_L is a subset of Dom R_L. The relation R_L is on a set U. C_L is termed the competence set of the LD relation L. A description of LD relations and a discussion of their properties is in chapter two of [Par:01].

The LD relations may be used to describe programs. The relation component of the LD relation L describes a set of states such that if execution starts in state x it may terminate in state y. The set U is the set of states. The competence set of L is such that if execution starts in a state that is in the competence set then its termination is guaranteed.

2.4 Functions

A function $f:A \rightarrow B$ is a special relation such that for each element $a \in A$ there is exactly one element $b \in B$. This is written as $f(a) = b$. The domain of the function (denoted by **dom** f) is the set of values in A for which the function is defined. The domain of the function is A provided that f is a total function. The co-domain of the function is B. The range of the function (denoted **rng** f) is a subset of the co-domain and consists of:

$$\textbf{rng } f = \{r \mid r \in B \text{ such that } f(a) = r \text{ for some } a \in A\}.$$

Functions may be partial or total. A partial function may be undefined for values of A. Total functions are defined for every value in A. Functions are an essential part of mathematics and computer science, and examples of useful functions include the trigonometric functions, logarithmic, and exponential.

Partial functions arise in computer science as a program may be undefined or fail to terminate for some values of its arguments. Care is required to ensure that the partial function is defined for the argument to which it is to be applied. Total functions are defined everywhere for all of their arguments. Two functions f and g are equal if:

1. dom f = dom g
2. $f(a) = g(a)$ for all $a \in$ dom f.

A function f is less defined than a function g ($f \subseteq g$) if the domain of f is a subset of the domain of g and the functions agree for every value on the domain of f:

1. dom $f \subseteq$ dom g
2. $f(a) = g(a)$ for all $a \in$ dom f.

The composition of functions is similar to the composition of relations. Suppose f: A $\rightarrow$B and g: B $\rightarrow$C then $g \circ f$: A $\rightarrow$C is a function and for $x \in$ A this is written as $g \circ f(x)$ or $g(f(x))$. Consider the function f: R $\rightarrow$ R such that $f(x) = x^2$ and the function g:R$\rightarrow$R such that $g(x) = x+2$. Then $g \circ f(x) = g(x^2) = x^2 + 2$. However, $f \circ g(x) = f(x+2) = (x+2)^2 \neq x^2 + 2$.

Hence, composition of functions is not commutative. The composition of functions is associative, for example, consider f: A $\rightarrow$B, g: B $\rightarrow$C, and h: C$\rightarrow$D, then

$$h \circ (g \circ f) = (h \circ g) \circ f.$$

A function f: A$\rightarrow$B is injective (one to one) if $f(a_1) = f(a_2) \Rightarrow a_1 = a_2$. For example, consider the function f: R $\rightarrow$R with $f(x) = x^2$. Then $f(3) = f(-3) = 9$ and so f is not one to one.

A function f: A$\rightarrow$B is surjective (onto) if given any $b \in$ B there exists $a \in$ A such that $f(a) = b$. Consider the function f: R $\rightarrow$R with $f(x) = x+1$. Clearly, given any $r \in$ R then $f(r-1) = r$ and so f is onto. A function is bijective if it is one to one and onto.

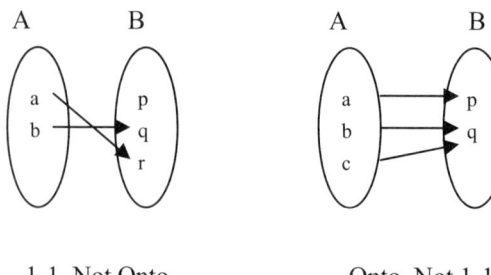

1-1, Not Onto   Onto, Not 1-1

The inverse of a relation was discussed in section 2.4 and the relational inverse of a function f: A →B clearly exists. The relational inverse of the function may or may not be a function. However, if the relational inverse is a function it is denoted by f^{1}: B →A. A total function has an inverse if and only if it is bijective.

The identity function 1_A : A →A is a function such that $1_A(a) = a$ for all $a \in$ A. Clearly, when the inverse of the function exists then $f^{1} \circ f = 1_A$ and $f \circ f^{1} = 1_B$. Further information on sets, relations, and functions is available in [Pif:91].

2.5 Logic

The first formal logic (syllogistic logic) in Western civilization was invented by Aristotle[2] [Ack:94] in the 4th century B.C. This logic could handle terms only and these include the subject and predicate of propositions. The propositional calculus allows a truth-value to be associated with each proposition and is widely employed in mathematics and logic. There are a rich set of connectives employed in the calculus for truth functional operations, and these include $A \Rightarrow B$, $A \wedge B$, $A \vee B$ which denote, respectively, the conditional if A then B, the conjunction of A and B, and the disjunction of A and B. A truth table of the logical operations may be constructed, and this details the truth values that result from the operation depending on the Boolean values of the constituent propositions. Propositional logic is the classical two valued logic and there are two possible Boolean values (i.e., *true* and *false*). There are other logics also: for example, the 3-valued logics, in which the propositions may have three truth-values, namely true, false, and undefined. Undefined values are discussed further in Chapter 3.

The propositional calculus is widely employed in computer science. Every program uses logic extensively: for example, the Boolean condition in an if then else statement determines whether a particular statement will be executed

---

[2] Aristotle was a Greek philosopher. He was a pupil of Plato but later founded his own school of philosophy. His writings on philosophy are extensive including metaphysics, epistemology, aesthetics, logic, ethics, biology, and politics.

or not; similarly, the Boolean condition in a while statement determines whether there will be another iteration of the loop.

There are many well-known properties of the propositional calculus. These include properties such as $A \wedge F = F$ and $A \wedge T = A$; $A \vee T = T$ and $A \vee F = A$. The $\wedge$ and $\vee$ operators are idempotent; i.e., the following properties are true:

$$A \wedge A = A.$$
$$A \vee A = A.$$

The $\wedge$ operator distributes over the $\vee$ operator and vice-versa:

$$A \wedge (B \vee C) = (A \wedge B) \vee (A \wedge C).$$
$$A \vee (B \wedge C) = (A \vee B) \wedge (A \vee C).$$

The $\wedge$ and $\vee$ operators are associative:

$$A \wedge (B \wedge C) = (A \wedge B) \wedge C.$$
$$A \vee (B \vee C) = (A \vee B) \vee C.$$

De Morgans Law holds for the $\wedge$ and $\vee$ operators:

$$\neg(A \vee B) = \neg A \wedge \neg B.$$
$$\neg(A \wedge B) = \neg A \vee \neg B.$$

A formula in predicate calculus is built up from the basic symbols of the language; these symbols include variables; predicate symbols, including equality; function symbols, including the constants; logical symbols, e.g., $\exists$, $\wedge$, $\vee$, $\neg$, etc.; and the punctuation symbols, e.g., brackets and commas. The formulae of predicate calculus are then built from terms, where a *term* is a key construct, and is defined recursively as a variable or individual constant or as some function containing terms as arguments. A formula may be an atomic formula or built from other formulae via the logical symbols. Other logical symbols are then defined as abbreviations of the basic logical symbols.

The predicate calculus is widely employed in computer science. Often, the definition of a function is piecewise where the domain of the function is partitioned and each partition is given by constraints expressed in predicate logic.

One problem that arises is how to deal with undefined terms that may occur in predicate calculus expressions. Jones [Jon:90] has proposed the logic of partial functions, which is a 3-valued logic. Parnas has developed a theory that deals with undefinedness using a classical 2-valued logic. Dijkstra's **cand** and **cor** operators also deal with undefinedness. Undefinedness is discussed further in Chapter 3.

2.6 Tabular Expressions

Tables of constants have been used for millennia to define mathematical functions (e.g., the Plimpton[3] 322 cuneiform tablet from the Babylonians c. 1900 B.C.) Such tables[4] are used to present data in an organized and more easily referenced form.

Tabular expressions are a generalization of tables in which constants can be replaced by more general mathematical expressions. Conventional mathematical expressions are a special case of tabular expressions. Tabular expressions can represent sets, relations, functions, and predicates. In fact, everything that can be expressed as a tabular expression can be represented by a conventional expression. However, the advantage is that the tabular expression is usually easier to read and use than a conventional expression.[5] A complex conventional mathematical expression is replaced by a set of much simpler expressions. Tabular expressions have been applied to precisely document system requirements.

Consider a function $f(x,y)$ defined piecewise as follows:

$$f(x,y) = 0 \qquad \text{where } x \geq 0 \wedge y = 10;$$
$$f(x,y) = y^2 \qquad \text{where } x \geq 0 \wedge y > 10;$$
$$f(x,y) = -y^2 \qquad \text{where } x \geq 0 \wedge y < 10;$$
$$f(x,y) = x \qquad \text{where } x < 0 \wedge y = 10;$$
$$f(x,y) = x+y \qquad \text{where } x < 0 \wedge y > 10;$$
$$f(x,y) = x-y \qquad \text{where } x < 0 \wedge y < 10.$$

One problem with a definition in this form is that care is required to ensure that all cases are considered in the definition as it is easy to miss a case. Consider, an equivalent representation of the function represented as a tabular expression. Then it is very easy to verify that all cases are considered as this is clear from an examination of the headers of the table.

The evaluation of the function for a particular (x,y) involves determining the appropriate row and column from the headers of the table and computing the grid element for that row and column. For example, the evaluation of $f(2,3)$ involves the selection of row 1 of the grid (as $x = 2 \geq 0$ in H_1) and the selection of column 3 (as $y = 3 < 10$ in H_2). Hence the value of $f(2,3)$ is given by the expression in row 1 and column 3 of the grid: i.e., $-y^2$ evaluated with $y = 3$ resulting in -9. The table simplifies the definition of the function.

---

[3] The Plimpton 322 tablet was discovered in Iraq and now resides in Columbia University. It is a clay table with 15 rows and 4 columns. The entries in the table are in the Babylonian sexigesimally (base 60) notation and the table has been translated to Decimal notation. Further information is on http://www.math.ubc.ca/~cass/courses/m446-03/pl322/pl322.html.

[4] Tables are used extensively in today's society: e.g., bus and train timetables, examination timetables, logarithmic tables, sine and cosine tables, etc.

[5] I acknowledge that some of the tabular expressions are not for the faint-hearted.

	y = 10	y > 10	y < 10	H
x ≥ 0	0	y^2	$-y^2$	G
x < 0	x	x+y	x-y	

H_1 (left label)

Fig. 2.2. Tabular Expression (Normal Table)

2.7 Probability and Applied Statistics

The probability of an event occurring is an indication of how likely the event is to occur. Probability theory[6] provides a mathematical indication of the likelihood of the event occurring and the mathematical probabilities range between 0 and 1. A probability of 0 indicates that the event cannot occur whereas a probability of 1 indicates that the event is guaranteed to occur. A probability value greater than 0.5 indicates that the event is more likely to occur than not to occur.

A sample space is the set of all possible outcomes of an experiment and an event E is a subset of the sample space. The probability of the union of disjoint events is the sum of their individual probabilities: i.e.,

$$P(\cup_{i=1}^{n} E_i) = \Sigma_{i=1}^{n} P(E_i).$$

The probability of the union of two events (not necessarily disjoint) is given by:

$$P(E \cup F) = P(E) + P(F) - P(E \cap F).[7]$$

The probability of an event E not occurring is denoted by E^c and is given by $1 - P(E)$. The probability of an event E occurring given that an event F has occurred is termed conditional probability (denoted by $P(E|F)$) and is given by:

$$P(E|F) = P(EF) / P(F).$$

Bayes formula enables the probability of an event E to be determined by a weighted average of the conditional probability of E given that the event F occurred and the conditional probability of E given that F has not occurred:

$$P(E) = P(E|F)P(F) + P(E|F^c)P(F^c).$$

---

[6] The history of probability theory dates back to the 17th Century and arose out of a gambling dispute. The dispute involved Chevalier De Mére who drew Pascal's attention to an apparent contradiction in a roll of a pair of dice. Pascal and Fermat investigated further and the theory of probability was born.

[7] E ∩ F is denoted by EF.

Two events E, F are independent if knowledge that F has occurred does not change the probability that E has occurred. Two events E, F are independent if:

$$P(EF) = P(E)P(F).$$

Often, some numerical quantity determined by the result of the experiment is of interest rather than the result of the experiment itself. These numerical quantities are termed random variables. A random variable is termed discrete if it can take on a finite or countable number of values; otherwise it is termed continuous. The distribution function is the probability that the random variable X takes on a value less than or equal to x. It is given by:

$$F(x) = P(X \leq x).$$

The probability mass function $p(a)$ of X is given by:

$$p(a) = P\{X = a\}.$$

A random variable is continuous if there exists a function f such that:

$$P\{X \in B\} = \int_B f(x)\, dx.$$

The function f is termed the probability density function and $\frac{d}{da} F(a) = f(a)$.

The expected value (i.e., mean) of a discrete random variable X (denoted E[X]) is given by the weighted average of the possible values of X:

$$
\begin{aligned}
E[X] \quad &= \Sigma_i x_i\, P\{X = x_i\} &&\text{discrete random variable.} \\
&= \int_{-\infty}^{\infty} xf(x)\, dx &&\text{continuous random variable.}
\end{aligned}
$$

Further, $E[g(X)] = \Sigma_i g(x_i)P\{X = x_i\}$ and $\int_{-\infty}^{\infty} g(x)f(x)\, dx$ for the discrete and continuous case respectively. The variance of a random variable (i.e., spread of values from mean) is given by:

$$Var(X) = E[X^2] - (E[X])^2.$$

The standard deviation (σ) is given by $\sqrt{Var(X)}$. The covariance of two random variables X,Y is given by:

$$Cov(X,Y) = E[XY] - E[X]E[Y].$$

A positive covariance $(\text{Cov}(X,Y) \geq 0)$ indicates that Y tends to increase as X does whereas a negative covariance indicates that Y tends to decrease as X increases. The correlation of two random variables $(\text{Corr}(X,Y)$[8]$)$ is given by:

$$\text{Corr}(X,Y) = \text{Cov}(X,Y) / \sqrt{(\text{Var}(X)\text{Var}(Y))}.$$

There are a number of special random variables and these include a Bernouilli trial where there are two possible outcomes from an experiment: i.e., success or failure:

$$P\{X = 0\} = 1\text{-}p.$$
$$P\{X = 1\} = p.$$

The mean is given by p and the variance by $p(1\text{-}p)$. The Binomial Distribution involves n Bernouilli trials, each of which results in success or fail:

$$P\{X = i\} = \binom{n}{i} p^i (1\text{-}p)^{n-i}$$

with the mean of the Bernouilli random variable given by np and the variance by $np(1\text{-}p)$.

The Poisson distribution may be used as an approximation for the Binomial when n is large and p is small. A Poisson random variable is given by:

$$P\{X = i\} = e^{-\lambda}\lambda^i / i!$$

with the mean and variance of the Poisson Distribution given by λ.

There are many other well-known random variables such as:

Distribution Name	Density Funtion	Mean / Variance
Hypergeometric	$P\{X = i\} = \binom{N}{i}\binom{M}{N-i} / \binom{N+M}{i}$;	
Uniform	$f(x) = 1/(\beta\text{-}\alpha) \; \alpha \leq x \leq \beta, 0$	$(\alpha+\beta)/2, \; (\beta\text{-}\alpha)^2/12$
Exponential	$f(x) = \lambda e^{-\lambda x}$	$1/\lambda, 1/\lambda^2$
Normal	$f(x) = 1 / \sqrt{2\pi}\sigma[\; e^{-(x-\mu)_2 /2\sigma_2}]$	μ, σ^2

Table 2.2. Probability Distributions

A good account of probability and statistics is in [Ros:87]. Probability theory has been applied to develop software reliability predictors of the mean time to failure (MTTF) or the mean time between failure (MTBF).

---

[8] The correlation of two random variables has a value between ± 1.

2.8 Calculus

Calculus was discovered independently by Newton and Leibnitz in the 17th century.[9] This section provides a brief introduction to some of the essential concepts in Calculus. The concept of the limit of a function is fundamental and the limit of a function f at $x = a$ (denoted by $\lim_{x \to a} f(x)$) is said to be l if given $\varepsilon > 0$ there exists $\delta > 0$ such that $|f(x) - l| < \varepsilon$ when $|x - a| < \delta$. The limit of a function exists if and only if the left hand and right hand limit of the function exists. The left hand limit is denoted by $\lim_{x \to a-} f(x)$ and the right hand limit is given by $\lim_{x \to a+} f(x)$.

The function f is continuous at $x = a$ if the limit of the function f exists and is equal to $f(a)$: i.e., given $\varepsilon > 0$ there exists $\delta > 0$ such that $|f(x) - f(a)| < \varepsilon$ when $|x - a| < \delta$. The derivative of a function $y = f(x)$ is given by:

$$\lim_{h \to 0} f(x+h) - f(x) / h.$$

The derivative of a function $y = f(x)$ is denoted by $\frac{dy}{dx}$ or by $f'(x)$. A function f that has a derivative at $x = a$ is continuous at a. The interpretation of the derivative of a real valued continuous function f at $x = a$ is the slope of the curve at $x = a$. There are many well-known rules of derivatives:

$$\frac{d}{dx} (f(x) + g(x)) = f'(x) + g'(x).$$
$$\frac{d}{dx} (f(x) * g(x)) = f'(x)*g(x) + f(x)*g'(x).$$
$$\frac{d}{dx} (f \circ g(x)) = f'(g(x))*g'(x).$$

The derivative of a constant is zero: i.e., $\frac{d}{dx} k = 0$; the derivative of a function that is a power of x is given by: $\frac{d}{dx} x^n = nx^{n-1}$; the derivative of the sine function is: $\frac{d}{dx} \operatorname{Sin} x = \operatorname{Cos} x$; and the derivative of the cosine function is: $\frac{d}{dx} \operatorname{Cos} x = -\operatorname{Sin} x$.

The derivative of the function f is positive when f is an increasing function and negative when f is a decreasing function. A function attains a local maximum when $\frac{dy}{dx} = 0$ and $d^2y/dx^2 < 0$. It attains a local minimum when $\frac{dy}{dx} = 0$ and $d^2y/dx^2 > 0$.

The integral of a function is essentially the inverse of the derivative of the function: i.e., $\int f'(x)dx = f(x) + k$. There are many well-known properties of the integral operator:

$$\int (f(x) + g(x))dx = \int f(x)\, dx + \int g(x)\, dx.$$
$$\int f g'(x)dx = f(x)g(x) + \int f'(x)g(x)dx.$$

---

[9] There is some evidence that Indian mathematicians in Kerala in India were already familiar with Calculus 300 years before Newton and Leibnitz.

The interpretation of the integral $\int_a^b f(x)dx$ is the area under the curve $f(x)$ between a and b. The following are well-known examples:

$$\int x^n \, dx = x^{n+1}/(n+1) + k.$$
$$\int Sin \, x \, dx = - Cos \, x + k.$$
$$\int Cos \, x \, dx = Sin \, x + k.$$

Many well-known functions (e.g., the Trigonometric sine and cosine functions) can be expressed as a Taylor power series of x. The Taylor series of a function f about $x = a$ is given by:

$$f(x) = f(a) + (x-a)f'(a) + \frac{1}{2}(x-a)^2!f''(a) + \ldots\ldots + \frac{1}{n!}(x-a)^n f^n(a) + \ldots$$

The Taylor series requires the function to be infinitely differentiable and the convergence of the power series needs to be considered.

The application of calculus is extensive in physics and engineering. Space does not permit a more detailed description and the reader is referred to texts (e.g., Boa:66] that describe the richness of calculus including ordinary and partial differential equations; Fourier Series; and Laplace transforms.

2.9 Matrix Theory

Matrices[10] arose in practice as a means of solving a set of linear equations. For example, consider the set of equations:

$$ax + by = r.$$
$$cx + dy = s.$$

Then the coefficients of the linear equations may be represented by the matrix A $= \left(\begin{smallmatrix} a & b \\ c & d \end{smallmatrix}\right)$ and the equations may be represented as the multiplication of a matrix A and a vector $\underline{x} = \left(\begin{smallmatrix} x \\ y \end{smallmatrix}\right)$ resulting in a vector $\underline{v} = \left(\begin{smallmatrix} r \\ s \end{smallmatrix}\right)$: i.e., $A\underline{x} = \underline{v}$. The vector $\underline{x}$ may be calculated by determining the inverse of the matrix A (provided that the inverse of A exists) and calculating $\underline{x} = A^{-1}\underline{v}$.

The inverse of a matrix A exists when its determinant is non-zero and the determinant of a 2x2 matrix A $= \left(\begin{smallmatrix} a & b \\ c & d \end{smallmatrix}\right)$ is given by det A $= ad-cb$. The transpose of A (denoted by A^T) is given by $A^T = \left(\begin{smallmatrix} a & c \\ b & d \end{smallmatrix}\right)$. The inverse of the matrix A (denoted by A^{-1}) is given by $\frac{1}{\det A}\left(\begin{smallmatrix} d & -b \\ -c & a \end{smallmatrix}\right)$ and A. $A^{-1} = A^{-1}.A = I$. The matrix I is the identity matrix of the algebra of 2x2 matrices and is given by $I = \left(\begin{smallmatrix} 1 & 0 \\ 0 & 1 \end{smallmatrix}\right)$.

---

[10] The first reference to matrices and determinants appears in Chinese mathematics c. 200 B.C.

The matrix representation of the linear equations and the solution of the linear equations are as follows:

$$\begin{vmatrix} a & b \\ c & d \end{vmatrix} \begin{vmatrix} x \\ y \end{vmatrix} = \begin{vmatrix} r \\ s \end{vmatrix}$$

$$\begin{vmatrix} x \\ y \end{vmatrix} = \begin{vmatrix} {}^d/_{\det A} & {}^{-b}/_{\det A} \\ {}^{-c}/_{\det A} & {}^a/_{\det A} \end{vmatrix} \begin{vmatrix} r \\ s \end{vmatrix}$$

The addition of two 2x2 matrices $A = \begin{pmatrix} a & b \\ c & d \end{pmatrix}$ and $B = \begin{pmatrix} p & q \\ q & s \end{pmatrix}$ is given by a matrix whose entries is the addition of the individual components of A and B; the multiplication of two 2x2 matrices is given by:

$$AB = \begin{vmatrix} ap+br & aq+bs \\ cp+dr & cq+ds \end{vmatrix}$$

The addition of two matrices is commutative: i.e., $A + B = B + A$; the multiplication of matrices is not commutative: i.e., $AB \neq BA$. Matrices are an example of an algebraic structure known as the ring.[11]

More general sets of linear equations may be solved by $m \times n$ matrices (i.e., a matrix with m rows and n columns). The multiplication of two matrices A and B is meaningful if and only if the number of columns of A is equal to the number of rows of B: i.e., A is an $m \times n$ matrix and B is a $n \times p$ matrix and the resulting matrix AB is a $m \times p$ matrix.

Let $A = (a_{ij})$ where i ranges from 1 to m and j ranges from 1 to n; $B = (b_{jl})$ where j ranges from 1 to n and l ranges from 1 to p. Then AB is given by (c_{il}) where i ranges from 1 to m and l ranges from 1 to p and c_{il} is given by:

$$c_{il} = \sum_{k=1}^{n} a_{ik} b_{kl}.$$

The identity matrix I is a $n \times n$ matrix and the entries are given by r_{ij} where $r_{ii} = 1$ and $r_{ij} = 0$ where $i \neq j$. The inverse of a $n \times n$ matrix exists if and only if its determinant is non-zero. The reader is referred to texts on matrix theory for more detailed information.

2.10 Finite State Machines

A finite state machine (also known as finite state automata) is a quintuple (Σ, Q, δ, q_0, T). The alphabet of the finite state machine (FSM) is given by Σ; the set of

---

[11] A ring (R,+,*) is essentially a structure with two binary operations such that (R,+) is a commutative group; (R, *) is a semi-group and the left and right distributive properties of multiplication over addition hold. For further information see [Her:75].

states is given by Q; the transition function is defined by $\delta : Q \times \Sigma \rightarrow Q$; the initial state is given by q_0; and the set of accepting states is given by T where $T \subseteq Q$. A string is given by a sequence of alphabet symbols: i.e., $s \in \Sigma^*$ and the transition function δ can be extended to $\delta^* : Q \times \Sigma^* \rightarrow Q$.

A string $s \in \Sigma^*$ is accepted by the finite state machine if $\delta^*(q_0, s) = q_t$ where $q_t \in T$. A finite state machine is termed deterministic if the transition function δ is a function; otherwise it is termed nondeterministic. A nondeterministic automata is one for which the next state is not uniquely determined from the present state and input.

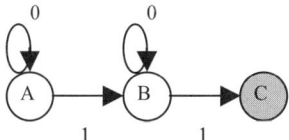

Fig. 2.3. Finite State Machine

For the example above the input alphabet is given by $\Sigma = \{0,1\}$; the set of states by $\{A,B,C\}$; the start state by A; the final state by $\{C\}$; and the transition function is given in the table below. The language accepted by the automata is the set of all binary strings that end with a 1 that contain exactly two 1s.

State	0	1
A	A	B
B	B	C
C	-	-

Table 2.3.. State Transition Table

The set of strings (or language) accepted by an automaton M is denoted *L(M)*. A language is termed regular if it is accepted by some finite state machine. Regular sets are closed under union, intersection, concatenation, complement, transitive closure. That is, for regular sets $A,B \subseteq \Sigma^*$ then:

- $A \cup B$ and $A \cap B$ are regular.
- $\Sigma^* \setminus A$ (i.e., A^c) is regular.
- AB and A* is regular.

The proof of these properties is demonstrated by constructing finite state machines to accept these languages. The proof for $A \cap B$ is to construct a machine $M_{A \cap B}$ that mimics the execution of M_A and M_B and is in a final state if and only if both M_A and M_B are in a final state. Finite state machines are useful in designing systems that process sequences of data.

2.11 Graph Theory

A graph G is a pair of finite sets (V,E) such that E is a binary symmetric relation on V. The set V is termed the set of vertices (or nodes) and the set E is the set of edges. A directed graph is a pair of finite sets (V,E) where E is a binary relation that may not be symmetric. An edge $e \in E$ consists of a pair $<x,y>$ where x, y are adjacent nodes in the graph. The degree of x is the number of nodes that are adjacent to x. The set of edges is denoted by E(G) and the set of vertices is denoted by V(G).

The example below is of a directed graph with three edges and four vertices.

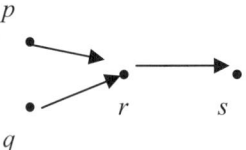

A graph G' = (V', E') is a subgraph of G if $V' \subseteq V$ and $E' \subseteq E$. A weighted graph is a graph G = (V,E) together with a weighting function w:E $\rightarrow N$ which associates a weight with every edge in the graph. For example, the weight of an edge may be applied to model the bandwidth of a telecommunications link between two nodes.

For an adirected graph the weight of the edge is the same in both directions: i.e., $w(v_i,v_j) = w(v_j,v_i)$ for all edges $<v_i,v_j>$ in the graph. The degree of a vertex v is the number of distinct edges incident to v. That is, deg $v = |\{u \in V: vu \in E\}|$. A vertex of degree 0 is called an isolated vertex. Two vertices x,y are adjacent if $xy \in E$, and x and y are said to be incident to the edge xy.

A graph G = (V,E) is said to be complete if all the vertices are adjacent: i.e., $E = V \times V$. A path $v_1, v_2, ... ,v_k$ from vertex v_1 to v_k is of length $k-1$ and consists of the sequence of edges $< v_1, v_2 >,< v_2, v_3 >,...< v_{k-1}, v_k >$. The vertices in the path are all distinct apart from possibly v_1 and v_k. The path is said to be a cycle if $v_1 = v_k$. A Hamiltonian[12] path in a graph G is a path that visits every vertex once and once only. It is applicable to the traveling salesman problem where a salesman wishes to travel to k cities in the country without visiting any city more than once.

A graph is said to be connected if for any two given vertices v_1, v_2 in V there is a path from v_1 to v_2. A connected graph with no cycles is a tree. A directed acylic graph (dag) is a directed graph that has no cycles.

Two graphs G = (V,E) and G' = (V',E') are said to be isomorphic if there exists a bijection f: V $\rightarrow$ V' such that for any $u,v \in V$, $uv \in E$ if and only if

---

[12] Sir Rowan Hamilton was an Irish mathematician and astronomer royal in Ireland. He discovered the Quaternions (a generalization of complex numbers that have applications in robotics). He was a contemporary of George Boole, first professor of mathematics at Queens College Cork, Ireland and inventor of Boolean Algebra.

$f(u)f(v) \in E'$. The mapping f is called an isomorphism. Two graphs that are isomorphic are equivalent apart from a relabeling of the nodes and edges.

2.12 Tools for Mathematics

There have been tools available for mathematics for several millennia. For example, the abacus[13] is a portable tool for counting that is built out of wood and beads and has been available since 3000 B.C. More modern tools have been introduced over time and these include tools such as the slide rule and electronic calculators to perform calculation.

Mathematica is one of the most popular tools available for mathematics and may be employed to do computations and to perform 2-dimensional or 3-dimensional plots of mathematical functions. It allows integrals to be computed and allows matrix operations to be performed. It is a programming language and the user may define powerful functions as appropriate.

It has been applied to many scientific areas including astronomy and statistics, and it is very effective in working with large numbers. It can also represent real numbers precisely. Further information on Mathematica is available from Wolfram Research http://www.wolfram.com/.

2.13 Summary

This chapter provided a brief introduction to the mathematics essential for software engineering. Software engineers need a firm foundation in mathematics in order to be engineers in the classical sense. The use of mathematics will enable the software engineer to produce high-quality products that are safe to use. The mathematics required by software engineers includes set theory, relations, functions, mathematical logic, tabular expressions, matrix theory, graph theory, finite state automata, calculus and probability theory.

The emphasis is on mathematics that can be applied rather than mathematics for its own sake. The engineering approach aims to show how mathematics can be used to solve practical problems rather than an emphasis on the proof of mathematical correctness. The engineer applies mathematics and models to the design of the product and the analysis of the design is a mathematical activity.

The advantage of mathematics is that it allows rigorous analysis and avoids an overreliance on intuition. Mathematics provides precise unambiguous statements and the mathematical proof of a theorem provides a high degree of confidence in its correctness. The objective of the teaching of mathematics for

---

[13] The abacus was invented by the Chinese c. 3000 B.C.

software engineering is to teach students on using and applying mathematics to program well and to solve practical problems.

Many programming courses treat the mathematics of programming as if it is too difficult or irrelevant to the needs of computer science students. Instead, many programming courses often focus on teaching the latest programming language rather than placing the emphasis on the design of useful products. It is important to teach students problem-solving skills such as formulating the problem; decomposing a problem into smaller problems; and integrating the solution. This mathematical training will provide a solid foundation for the student to enable practical industrial problems to be solved.

3
Logic for Software Engineering

3.1 Introduction

Logic is concerned with reasoning and the validity of arguments. It allows conclusions to be deduced from premises according to logical rules that ensure the truth of the conclusion provided that the premises are true. Logic plays a key role in mathematics but is regarded as a separate discipline from mathematics. The 20th century witnessed an attempt to show that all mathematics can be derived from formal logic. However, this effort failed as it was discovered that there were truths in the formal system of arithmetic that could not be proved within the system. This was proved by the Austrian logician Kurt Goedel.[1]

The roots of logic are in antiquity and early work was done by Aristotle in a work known as the *Organon* [Ack:94] in the 4th century B.C. Aristotle regarded logic as a useful tool of inquiry into any subject. The formal logic developed by Aristotle is known as *syllogistic logic*, where a syllogism is a form of reasoning in which a conclusion is drawn from two premises. A common or middle term is present in the two premises but not in the conclusion. Each premise that occurs in a syllogism is of a subject-predicate form. A well-known example of a syllogism is:

> All Greeks are mortal.
> Socrates is a Greek.
> ------------------------
> Therefore Socrates is mortal.

The common (or middle) term in this example is Greek and the argument is valid. Aristotle identified and classified the various types of syllogistic arguments and determined those that were valid or invalid. The invention of syllogistic logic was an impressive achievement as it was the first attempt at a science of logic. There are many limitations to what the syllogistic logic may

---

[1] This result is known as Goedel's Incompletness Theorem and was proved in 1929.

express and on its suitability as a representation of how the mind works. There are four types of premises in a syllogistic argument either affirmative or negative and either universal or particular.

Type	Symbol	Example
Universal affirmative	G A M	All Greeks are mortal.
Universal negative	G E M	No Greek is mortal.
Particular affirmative	G I M	Some Greek is mortal.
Particular negative	G O M	Some Greek is not mortal.

Table 3.1. Types of Syllogistic Premises

Each premise contains a subject and a predicate, and the middle term may act as subject or predicate. This leads to four basic forms of syllogistic arguments where the middle is the subject of both premises; where the middle is the predicate of both premises; and where the middle is the subject of one premise and the predicate of the other premise.

(i)	(ii)	(iii)	(iv)
M P	P M	P M	M P
M S	S M	M S	S M
-----	----	-----	------
S P	S P	S P	S P

There are four types of premises (A, E, I, O) and therefore sixteen sets of premise pairs for each of the forms above. However, only some of these premise pairs will yield a valid conclusion. Aristotle went through every possible premise pair to determine if a valid argument may be derived. He employs rules of deduction to prove the validity of the conclusions. The syllogistic argument above is of form (iv) and is valid:

$$G A M$$
$$S I G$$
$$\text{-------}$$
$$S I M$$

Aristotle's syllogistic logic is a *term-logic* rather than a propositional logic. He uses letters to stand for the individual terms in a proposition whereas in propositional logic a letter stands for a complete proposition. Aristotle also did work in identifying and classifying bad arguments (known as fallacies) and the names that he identified are still in use today.

Propositional logic was later discovered by Chrysippus (third head of the Stoics) in the 3rd century B.C., but it did not replace Aristotle's syllogistic logic. Propositional logic was re-discovered in the 19th century and it is discussed in the next section.

3.2 Propositional Logic

Propositional logic is the study of propositions, a statement that is either true or false. An example of a proposition P is given by the statement Today is Wednesday. Then this statement is either true or false,[2] and is true if today is Wednesday and false otherwise. A propositional variable is used to stand for a proposition (e.g., let the variable P stand for the proposition '$2 + 2 = 4$' which is true), and a propositional variable therefore takes the value true or false. The negation of a proposition P (denoted ¬P) is the proposition that is true if and only if P is false, and is false if and only if P is true. A formula in propositional calculus may contain several propositional variables, and the truth or falsity of the individual variables needs to be known prior to determining the truth or falsity of the logical formula.

Each propositional variable has two possible values, and a formula with n-propositional variables has 2^n values associated with the propositional variables. The set of values associated with the n variables may be used to derive a truth table with 2^n rows and $n + 1$ columns. Each row gives each of the 2^n values that the n variables may take and column $n + 1$ gives the result of the logical expression for that set of values of the propositional variables.

A rich set of connectives is employed in propositional calculus to combine propositions and to build the well-formed formulae of the calculus. This includes the conjunction of two propositions ($A \land B$), the disjunction of two propositions ($A \lor B$), and the implication of two propositions ($A \Rightarrow B$). These connectives allow compound propositions to be formed, and the truth of the compound propositions is determined from the truth-values of the constituent propositions and the rules associated with the logical connective. The meaning of the logical connectives is given by truth tables.[3]

Propositional logic allows further truths to be derived by logical reasoning or rules of inference. These rules enable new propositions to be deduced from a set of existing propositions provided that the rules of inference for the logic are followed. A valid argument (or deduction) is truth preserving: i.e., if the set of propositions is true then the deduced proposition will also be true. The various logical rules include rules such as *modus ponens*: i.e., given the truth of the proposition A and the proposition $A \Rightarrow B$, then the truth of proposition B follows.

---

[2] Time zones are ignored and it is assumed that the statement is made with respect to a specific place and time.

[3] Basic truth tables were first used by Frege, and developed further by Post and Wittgenstein.

The propositional calculus is employed in reasoning about propositions and may be applied to formalize arguments in natural language. It has also been applied to computer science and the term Boolean algebra is named after the English mathematician 'George Boole'. Boole was the first professor of mathematics at Queens College, Cork in the mid-19th century and he formalized the laws of propositional logic that are the foundation for modern computers. Boolean algebra is used widely in programs: for example, the Boolean condition in an if then else statement determines whether a particular statement will be executed or not; similarly, the Boolean condition in a while or for loop will determine if the statement in the body of the loop will be executed.

3.2.1 Truth Tables

Truth tables enable the truth-value of a compound proposition to be determined from its underlying propositions. Compound propositions are formed from binary connectives such as conjunction, disjunction, negation, implication, and equivalence.

The conjunction of A and B (denoted $A \wedge B$) is true if and only if both A and B are true, and is false in all other cases. The disjunction of two propositions A and B (denoted $A \vee B$) is true if at least one of A and B are true, and false in all other cases. The disjunction operator is known as the *inclusive or* operator as it is also true when both A and B are true; there is also an *exclusive or* operator that is true exactly when one of A or B is true, and is false otherwise.

A	B	$A \wedge B$
T	T	T
T	F	F
F	T	F
F	F	F

Table 3.2. Conjunction

A	B	$A \vee B$
T	T	T
T	F	T
F	T	T
F	F	F

Table 3.3. Disjunction

The implication operation (A $\Rightarrow$ B) is true if whenever A is true means that B is also true; and also whenever A is false. It is equivalent (as shown by a truth table) to $\neg A \vee B$. The equivalence operation (A $\equiv$ B) is true whenever both A and B are true, or whenever both A and B are false.

A	B	$A \Rightarrow B$
T	T	T
T	F	F
F	T	T
F	F	T

Table 3.4. Implication

A	B	$A \equiv B$
T	T	T
T	F	F
F	T	F
F	F	T

Table 3.5. Equivalence

The not operator ($\neg$) is a unary operator (i.e., it has one argument) and is such that $\neg A$ is true when A is false, and is false when A is true.

A	$\neg A$
T	F
F	T

Table 3.6. Not Operation

3.2.2 Properties of Propositional Calculus

There are many well-known properties of propositional calculus and these properties enable logic expressions to be simplified and ease the evaluation of a complex logical expression. These include the commutative, associative, and distributive properties. The commutative property holds for the conjunction and disjunction binary operators. This means that the order of the two propositions may be reversed without affecting the resulting truth-value: i.e.,

$$A \land B = B \land A.$$
$$A \lor B = B \lor A.$$

The associative property holds for the conjunction and disjunction operators. This means that order of evaluation of a subexpression does not affect the resulting truth-value: i.e.,

$$(A \land B) \land C = A \land (B \land C).$$
$$(A \lor B) \lor C = A \lor (B \lor C).$$

The conjunction operator distributes over the disjunction operator and vice versa.

$$A \land (B \lor C) = (A \land B) \lor (A \land C).$$
$$A \lor (B \land C) = (A \lor B) \land (A \lor C).$$

The result of the logical conjunction of two propositions is false if one of the propositions is false (irrespective of the value of the other proposition):

$$A \land F = F = F \land A.$$

The result of the logical disjunction of two propositions, where one of the propositions is known to be false is given by the truth-value of the other proposition. That is, the Boolean value "F" acts as the identity for the disjunction operation:

$$A \lor F = A = F \lor A.$$

The result of the logical conjunction of two propositions, where one of the propositions is known to be true, is given by the truth-value of the other proposition. That is, the Boolean value "T" acts as the identity for the conjunction operation:

$$A \wedge T = A = T \wedge A.$$

The $\wedge$ and $\vee$ operators are idempotent. That is, when the arguments of the conjunction or disjunction operator is the same proposition A the result of the operation is A. The idempotent property allows expressions to be simplified:

$$A \wedge A = A.$$
$$A \vee A = A.$$

The law of the excluded middle is a fundamental property of the propositional calculus. This states that a proposition A is either true or false: i.e., there is no third logical value:

$$A \vee \neg A.$$

De Morgan was contemporary with Boole in the 19th century and the following is known as De Morgan's Law. It enables logical expressions to be simplified:

$$\neg (A \wedge B) = \neg A \vee \neg B.$$
$$\neg (A \vee B) = \neg A \wedge \neg B.$$

A proposition that is true for all values of its constituent propositional variables is known as a tautology. An example of a tautology is $A \vee \neg A$. A proposition that is false for all values of its constituent propositional variables is known as a contradiction. An example of a contradiction is $A \wedge \neg A$.

3.2.3 Proof in Propositional Calculus

One of the very useful features of logic is that further truths may be derived from existing truths. These truths are derived from the currently known truths by rules of inference that are truth preserving. One of the key properties of the propositional calculus is that it is both complete and consistent. The completeness property means that all true propositions are deducible in the calculus, and the consistency property means that there is no formula A such that both A and $\neg A$ are deducible in the calculus.

An argument in propositional logic consists of a sequence of propositions that are the premises of the argument and a further proposition that is the conclusion of the argument. One elementary way to see if the argument is valid is to produce a truth table to determine if the conclusion is true whenever all of the premises are true. Consider a set of premises $P_1, P_2, \dots P_n$ and conclusion Q. Then to determine if the argument is valid using a truth table involves a column in the truth table for each premise P_1, P_2, ... P_n, and then to identify the rows in the truth table for which these premises are all true. The truth-value of the conclusion Q is examined in each of these rows, and if Q is true for each case for which P_1, P_2, ... P_n are all true then the argument is valid. This is equivalent to $P_1 \wedge P_2 \wedge ... \wedge P_n \Rightarrow Q$ is a tautology. An alternate approach with truth tables is to assume the negation of the desired conclusion (i.e., $\neg Q$) and to show that the premises and the negation of the conclusion result in a contradiction (i.e., $P_1 \wedge P_2 \wedge ... \wedge P_n \Rightarrow \neg Q$) is a contradiction. However, the use of truth tables becomes cumbersome when there are a large number of propositions involved, as there are 2^n truth table entries for n propositional variables.

Truth tables allow an informal approach to proof and the proof is provided in terms of the meanings of the propositions or logical connectives. The formalization of propositional logic is due to the German mathematician David Hilbert,[4] and it includes the definition of an alphabet of symbols and well-formed formulae of the calculus, the axioms of the calculus, and rules of inference for deduction.

The deduction of a new formulae Q is via a sequence of well-formed formulae P_1, P_2, ... P_n (where $P_n = Q$) such that each P_i is either an axiom, a hypothesis, or deducible from an earlier pair of formula P_j, P_k (where P_k is of the form $P_j \Rightarrow P_i$) and modus ponens. Modus ponens is a rule of inference that states that given propositions A, and $A \Rightarrow B$ then proposition B may be deduced. The deduction of a formula Q from a set of hypothesis H is denoted by $H \vdash Q$ and where Q is deducible from the axioms alone this is denoted by $\vdash Q$.

The deduction theorem states that if $H \cup \{P\} \vdash Q$ then $H \vdash P \Rightarrow Q$ and the converse of the theorem is also true: i.e., if $H \vdash P \Rightarrow Q$ then $H \cup \{P\} \vdash Q$. The Hilbert approach allows reasoning about symbols according to rules and to derive theorems from formulae irrespective of the meanings of symbols and formulae. However, the propositional calculus is sound; i.e., any theorem derived using the Hilbert approach is true; the calculus is also complete: i.e., every tautology has a proof (i.e., is a theorem in the formal system). The propositional calculus is consistent: i.e., it is not possible that both the well-formed formula A and $\neg A$ are deducible in the calculus. Propositional calculus is decidable: i.e., there is an algorithm to determine for any well-formed formula A whether A is a theorem of the formal system. The Hilbert style system is slightly cumbersome in conducting proof and is quite different from the normal use of logic in mathematical deduction.

---

[4] David Hilbert was a very influential German mathematician based at the University of Göttingen, Germany. He is well known for his work on invariant theory, analytic geometry, and functional analysis (Hilbert Spaces). He was a founder of the formalist school and launched a program that attempted to show that all mathematics was reducible to logic. Hilbert's objective was to deal with the crisis in foundations in mathematics in the late 19th/early 20th century following the paradoxes in set theory (e.g., Russel's paradox). The Hilbert program aimed to provide a system that was both consistent and complete; however, the program was dealt a fatal blow by the results of Kurt Goedel in 1929.

The German mathematician Gerhard Gentzen has developed a method known as *Natural Deduction,* and this is intended to be a formal system that is closer to natural reasoning. Natural induction includes rules for the introduction and elimination of the logical operators $\wedge$, $\vee$, $\Rightarrow$, $\equiv$, and also for reductio ab adsurdum.

There are ten inference rules in the Natural Deduction system and they are organized into two inference rules for each of the five logical operators $\wedge$, $\vee$, $\neg$, $\Rightarrow$ and $\equiv$.

The two inference rules per operator are the introduction rule and the elimination rule. The rules are defined as follows:

Rule	Definition	Description
$\wedge$ I	$$\frac{P_1,\ P_2,\ \dots\ P_n}{P_1 \wedge\ P_2 \wedge\ \dots \wedge P_n}$$	Given the truth of propositions P_1, P_2, ... P_n then the truth of the conjunction $P_1 \wedge\ P_2 \wedge\ \dots \wedge P_n$ follows. This rule shows how a conjunction can be introduced.
$\wedge$ E	$$\frac{P_1 \wedge\ P_2 \wedge\ \dots \wedge P_n}{P_i}$$	Given the truth the conjunction $P_1 \wedge\ P_2 \wedge\ \dots \wedge P_n$ then the truth of proposition P_i follows. This rule shows how a conjunction can be eliminated.
$\vee$ I	$$\frac{P_i}{P_1 \vee\ P_2 \vee\ \dots \vee P_n}$$	Given the truth of propositions P_i then the truth of the disjunction $P_1 \vee\ P_2 \vee\ \dots \vee P_n$ follows. This rule shows how a disjunction can be introduced.
$\vee$ E	$$\frac{P_1 \vee \dots \vee P_n, P_1 \Rightarrow E, \dots P_n \Rightarrow E}{E}$$	Given the truth of the disjunction $P_1 \vee\ P_2 \vee\ \dots \vee P_n$, and that each disjunct implies E, then the truth of E follows. This rule shows how a disjunction can be eliminated.
$\Rightarrow$ I	**From P_1, P_2, ... P_n infer P** $$\frac{}{(P_1 \wedge\ P_2 \wedge\ \dots \wedge P_n) \Rightarrow P}$$	This rule states that if we have a theorem that allows P to be inferred from the truth of premises P_1, P_2, ... P_n (or previously proved) then we can deduce $(P_1 \wedge\ P_2 \wedge\ \dots \wedge P_n) \Rightarrow P$. This is known as the Deduction Theorem.
$\Rightarrow$ E	$$\frac{P_i \Rightarrow P_j, P_i}{P_j}$$	This rule is known as modus ponens. The consequence of an implication follows if the antecedent is true (or has been previously proved).
$\equiv$ I	$$\frac{P_i \Rightarrow P_j, P_j \Rightarrow P_i}{P_i \equiv P_j}$$	If proposition P_i implies proposition P_j and vice versa then they are equivalent (i.e., $P_i \equiv P_j$).
$\equiv$ E	$$\frac{P_i \equiv P_j}{P_i \Rightarrow P_j, P_j \Rightarrow P_i}$$	If proposition P_i is equivalent to proposition P_j then proposition P_i implies proposition P_j and vice versa.

¬ I	**From P infer $P_1 \wedge \neg P_1$** $\overline{\quad\neg P\quad}$	If the proposition P allows a contradiction to be derived, then ¬P is deduced. This is an example of a proof by contradiction.
¬ E	**From ¬P infer $P_1 \wedge \neg P_1$** $\overline{\quad P\quad}$	If the proposition ¬P allows a contradiction to be derived, then P is deduced. This is an example of a proof by contradiction.

Table 3.7. Natural Deduction Rules

These rules are applied to derive further truths. Natural deduction is described in detail in [Gri:81, Kel:97].

3.2.4 Applications of Propositional Calculus

Propositional calculus may be employed to reasoning about arguments in natural language. First, the premises and conclusion of argument are identified and formalized into propositions. Propositional logic is then employed to determine if the conclusion is a valid deduction from the premises. Consider, for example, the following argument that aims to prove that Superman does not exist.

If Superman were able and willing to prevent evil, he would do so. If Superman were unable to prevent evil he would be impotent; if he were unwilling to prevent evil he would be malevolent; Superman does not prevent evil. If Superman exists he is neither malevolent nor impotent; therefore Superman does not exist.

First, letters are employed to represent the propositions as follows:

a : Superman is able to prevent evil
w : Superman is willing to prevent evil
i : Superman is impotent
m : Superman is malevolent
p : Superman prevents evil
e : Superman exists

Then, the argument above is formalized in propositional logic as follows:

$$P_1 \qquad (a \wedge w) \Rightarrow p$$
$$P_2 \qquad (\neg a \Rightarrow i) \wedge (\neg w \Rightarrow m)$$
$$P_3 \qquad \neg p$$
$$P_4 \qquad e \Rightarrow \neg i \wedge \neg m$$
$$P_1 \wedge P_2 \wedge P_3 \wedge P_4 \Rightarrow \neg e$$

The following is a proof that Superman does not exist using propositional logic.

1.	$\neg p$	P_3
2.	$\neg(a \wedge w) \vee p$	P_1 $(A \Rightarrow B \equiv \neg A \vee B)$
3.	$\neg(a \wedge w)$	$1,2$ $A \vee B, \neg B \vdash A$
4.	$\neg a \vee \neg w$	3, De Morgan's Law
5.	$(\neg a \Rightarrow i)$	P_2 ($\wedge$-Elimination)
6.	$\neg a \Rightarrow i \vee m$	$5, x \Rightarrow y \vdash x \Rightarrow y \vee z$
7.	$(\neg w \Rightarrow m)$	P_2 ($\wedge$-Elimination)
8.	$\neg w \Rightarrow i \vee m$	$7, x \Rightarrow y \vdash x \Rightarrow y \vee z$
9.	$(\neg a \vee \neg w) \Rightarrow (i \vee m)$	$8, x \Rightarrow z, y \Rightarrow z \vdash x \vee y \Rightarrow z$
10.	$(i \vee m)$	$4,9$ Modus Ponens
11.	$e \Rightarrow \neg(i \vee m)$	P_4 (De Morgan's Law)
12.	$\neg e \vee \neg (i \vee m)$	11, $(A \Rightarrow B \equiv \neg A \vee B)$
13.	$\neg e$	$10, 12$ $A \vee B, \neg B \vdash A$

Therefore, the conclusion that Superman does not exist is a valid deduction from the given premises.

3.2.5 Limitations of Propositional Calculus

The propositional calculus deals with propositions only. It is incapable of dealing with the syllogism "All Greeks are mortal; Socrates is a Greek; therefore Socrates is mortal" discussed earlier. This syllogism expressed in propositional calculus would be A, B therefore C where A stands for "All Greeks are mortal", B stands for "Socrates is a Greek", and C stands for "Socrates is mortal". Clearly, the argument is invalid in propositional logic. There is no way in propositional calculus to express the fact that all Greeks are mortal.

Predicate calculus deals with the limitations of propositional calculus by allowing variables and terms to be employed and using universal or existential quantification to express that a particular property is true of all (or at least one) values of a variable. Predicate calculus is discussed in the next section.

3.3 Predicate Calculus

Predicates are statements involving variables and these statements become propositions once these variables are assigned values. The set of values that the variables can take (universe of discourse) needs to be specified and the variables in the predicate take values from this universe. Predicate calculus enables expressions such as all members of the domain have a particular property: e.g., $(\forall x)Px$, or that there is at least one member that has a particular property: e.g., $(\exists x)Px$.

The syllogism "All Greeks are mortal; Socrates is a Greek; therefore Socrates is mortal" may be easily expressed in predicate calculus by:

$$(\forall x)(Gx \Rightarrow Mx)$$
$$Gs$$

$$Ms$$

In this example, the predicate Gx stands for x is a Greek and the predicate Mx stands for x is mortal. The formula $Gx \Rightarrow Mx$ states that if x is a Greek then x is mortal. The formula $(\forall x)(Gx \Rightarrow Mx)$ states for any x if x is a Greek then x is mortal. The formula Gs states that Socrates is a Greek and the formula Ms states that Socrates is mortal.

The predicate calculus is built from an alphabet of constants, variables, function letters, predicate letters, and logical connectives (including quantifiers). Terms are built from constants, variables, and function letters. A constant or variable is a term, and if $t_1, t_2, \ldots, t_k$ are terms, then $f_i^k(t_1, t_2, \ldots, t_k)$ is a term (where f_i^k is a k-ary function letter). Examples of terms include: π, x, $x^2 + y^2$, cos x where π is the constant 3.14159, x is a variable, $x^2 + y^2$ is shorthand for the function add(square(x), square(y)) where add is a 2-ary function letter and square is a 1-ary function letter.

The well-formed formulae are built from terms as follows. If P_i^k is a k-ary predicate letter, $t_1, t_2, \ldots, t_k$ are terms, then $P_i^k(t_1, t_2, \ldots, t_k)$ is a well-formed formula. If A and B are well-formed formulae then so are $\neg A$, A $\wedge$ B, A $\vee$ B, A $\Rightarrow$ B, A $\equiv$ B, $(\forall x)$A, and $(\exists x)$A. Examples of well-formed formulae include:

$$x = y,$$
$$(\forall x)(x > 2),$$
$$(\exists x)\, x^2 = 2, (\forall x)$$
$$(\exists y)\, x^2 = y.$$

The formula $x = y$ states that x is the same as y; the formula $(\forall x)(x > 2)$ states that every value of x is greater than the constant 2; $(\exists x)\, x^2 = 2$ states that there is an x such that the value of x is the square root of 2; $(\forall x)(\exists y)\, x^2 = y$ states that for every x there is a y such that the square of x is y.

The definition of terms and well-formed formulae specifies the syntax of the predicate calculus and the set of well-formed formulae gives the language of the predicate calculus. The terms and well-formed formulae are built from the symbols and these symbols are not given meaning in the formal definition of the syntax. The language defined by the calculus needs to be given an interpretation in order to give a meaning to the terms and formulae of the calculus. The interpretation needs to define the domain of values of the constants and variables, and to provide meaning to the function letters, the predicate letters, and the logical connectives.

The formalization of predicate calculus includes the definition of an alphabet of symbols (including constants and variables), the definition of function and predicate letters, logical connectives, and quantifiers. This leads to the definitions of the terms and well-formed formulae of the calculus. There are a set of axioms for predicate calculus and two rules of inference for the deduction of new formula from the existing axioms and previously deduced formulae. The deduction of new formulae Q is via a sequence of well-formed formulae P_1, P_2, ... P_n (where $P_n = Q$) such that each P_i is either an axiom, a hypothesis, or deducible from one or more of the earlier formulae in the sequence.

The two rules of inference are modus ponens and generalization. Modus ponens is a rule of inference that states that given predicate formulae A, and $A \Rightarrow B$ then the predicate formula B may be deduced. Generalization is a rule of inference such that given predicate formula A, then the predicate formula $(\forall x)$A may be deduced where x is any variable. The deduction of a formula Q from a set of hypothesis H is denoted by $H \vdash Q$ and where Q is deducible from the axioms alone this is denoted by $\vdash Q$. The deduction theorem states that if $H \cup \{P\} \vdash Q$ then $H \vdash P \Rightarrow Q$[5] and the converse of the theorem is also true: i.e., if $H \vdash P \Rightarrow Q$ then $H \cup \{P\} \vdash Q$. The Hilbert approach allows reasoning about symbols according to rules and to derive theorems from formulae irrespective of the meanings of symbols and formulae. However, the predicate calculus is sound: i.e., any theorem derived using the Hilbert approach is true, and the calculus is also complete.

The scope of the quantifier $(\forall x)$ in the well-formed formula $(\forall x)A$ is A. Similarly, the scope of the quantifier $(\exists x)$ in the well-formed formula $(\exists x)B$ is B. The variable x that occurs within the scope of the quantifier is said to be a bound variable. If a variable is not within the scope of a quantifier it is free. A well-formed formula is said to be closed if it has no free variables. A term t is free for x in A if no free occurrence of x occurs within the scope of a quantifier $(\forall y)$ or $(\exists y)$ in t. This means that the term t may be substituted for x without altering the interpretation of the well-formed formula A. The substitution therefore takes place only when no free variable in t will become bound by a quantifier in A through the substitution.

For example, suppose A is $\forall y\ (x^2+y^2 > 2)$ and the term t is y, then t is not free for x in A as the substitution of t for x in A will cause the free variable y in t to become bound by the quantifier $\forall y$ in A.

3.3.1 Properties of Predicate Calculus

An interpretation gives meaning to a formula and consists of a domain of discourse and a valuation function. If the formula is a sentence (i.e., does not contain any free variables) then the given interpretation of the formula is either true or false. If a formula has free variables, then the truth or falsity of the formula depends on the values given to the free variables. A free formula essentially describes a relation say, $R(x_1, x_2, ... x_n)$ such that $R(x_1, x_2, ... x_n)$ is true if $(x_1, x_2,$

---

[5] This should be stated more formally that if $H \cup \{P\} \vdash Q$ by a deduction containing no application of generalization to a variable that occurs free in P then $H \vdash P \Rightarrow Q$.

.... x_n) is in relation R. If a free formula is true irrespective of the values given to the free variables, then the formula is true in the interpretation.

A valuation (meaning) function gives meaning to the logical symbols and connectives. Thus associated with each constant c is a constant c_Σ in some universe of values Σ; with each function symbol f of arity k, we have a function symbol f_Σ in Σ and $f_\Sigma : \Sigma^k \rightarrow \Sigma$; and for each predicate symbol P of arity k a relation $P_\Sigma \subseteq \Sigma^k$. The valuation function, in effect, gives a semantics to the language of the predicate calculus L. The truth of a predicate P is then defined in terms of the meanings of the terms, the meanings of the functions, predicate symbols, and the normal meanings of the connectives.

Mendelson [Men:87] provides a technical definition of truth in terms of satisfaction (with respect to an interpretation M). Intuitively a formula F is *satisfiable* if it is *true* (in the intuitive sense) for some assignment of the free variables in the formula F. If a formula F is satisfied for every possible assignment to the free variables in F, then it is *true* (in the technical sense) for the interpretation M. An analogous definition is provided for *false* in the interpretation M.

A formula is *valid* if it is true in every interpretation; however, as there may be an uncountable number of interpretations, it may not be possible to check this requirement in practice. M is said to be a model for a set of formulae if and only if every formula is true in M.

There is a distinction between proof theoretic and model theoretic approaches in predicate calculus. *Proof theoretic* is essentially syntactic, and we have a list of axioms with rules of inference. In this way the theorems of the calculus may be logically derived and thus we may logically derive (i.e., $\vdash A$) the theorems of the calculus. In essence the logical truths are as a result of the syntax or form of the formulae, rather than the *meaning* of the formulae. *Model theoretical*, in contrast is essentially semantic. The truths derive essentially from the meaning of the symbols and connectives, rather than the logical structure of the formulae. This is written as $\vdash_M A$.

A calculus is *sound* if all the logically valid theorems are true in the interpretation, i.e., proof theoretic $\Rightarrow$ model theoretic. A calculus is complete if all the truths in an interpretation are provable in the calculus, i.e., model theoretic $\Rightarrow$ proof theoretic. A calculus is *consistent* if there is no formula A such that $\vdash A$ and $\vdash \neg A$. The predicate calculus is sound, complete, and consistent. Predicate calculus is not decidable: i.e., there is no algorithm to determine for any well-formed formula A whether A is a theorem of the formal system. The undecidability of the predicate calculus may be demonstrated by showing that if the predicate calculus is decidable then the halting problem (of Turing machines) is solvable.

3.3.2 Applications of Predicate Calculus

The predicate calculus is applicable to computer science and may be employed to formally state the system requirements of a proposed system. It may also be employed to define *f(x,y)* where *f(x,y)* is a piecewise defined function:

$$f(x,y) = x^2 - y^2 \quad \text{where } x \le 0 \wedge y < 0;$$
$$f(x,y) = x^2 + y^2 \quad \text{where } x > 0 \wedge y < 0;$$
$$f(x,y) = x + y \quad \text{where } x \ge 0 \wedge y = 0;$$
$$f(x,y) = x - y \quad \text{where } x < 0 \wedge y = 0;$$
$$f(x,y) = x + y \quad \text{where } x \le 0 \wedge y > 0;$$
$$f(x,y) = x^2 + y^2 \quad \text{where } x > 0 \wedge y > 0.$$

The predicate calculus allows rigorous proof to take place to verify the presence or absence of certain properties in a specification.

3.4 Undefined Values

Total functions $f: X \rightarrow Y$ are functions that are defined for every element in their domain and are the norm in mathematics. However, there are exceptions, for example, the function $y = 1/x$ is undefined at $x = 0$. Partial functions are quite common in computer science, and such functions may fail to be defined for one or more values in their domain. One approach to deal with this is to employ a precondition. The precondition limits the application of the function only to the restricted members of the domain for which the function is defined. This makes it possible to define a new set (a proper subset of the domain of the function) for which the function is total over the new set.

However, in practice undefined terms often arise[6] and therefore need to be dealt with. Consider the following example take from [Par: 93] where $\sqrt{x}$ is a function whose domain is the positive real numbers. Then the following expression is undefined:

$$((x > 0) \wedge (y = \sqrt{x})) \vee ((x \le 0) \wedge (y = \sqrt{-x})).$$

This is since the usual rules for evaluating such an expression involves evaluating each subexpression and then performing the Boolean operations. However, when $x \le 0$ the subexpression $y = \sqrt{x}$ is undefined; whereas when $x > 0$ the subexpression $y = \sqrt{-x}$ is undefined. Clearly, it is desirable that such expressions be handled, and that for the example above, the expression would evaluate to true. Classical 2-valued logic does not handle this situation adequately. There have been several proposals to deal with undefined values. These include Dijkstra's **cand** and **cor** operators in which the value of the left-hand operand determines whether the right hand operand expression is evaluated or not. The logic of partial functions [Jon:90] uses a 3-valued logic[7] and is discussed in the

---

[6] It is best to avoid undefined terms and expressions by taking care with the definitions.

[7] The above expression would evaluate to true under Jones 3-valued logic of partial functions.

next section. The approach of Parnas[8] is to employ a 2-valued logic that deals with undefinedness.

3.4.1 Logic of Partial Functions

Jones [Jon:90] has proposed the logic of partial functions (LPFs) as an approach to deal with terms that may be undefined. This is a 3-valued logic and a logical term may be true, false or undefined (denoted ⊥). The truth functional operators in classical 2-valued logic may be applied in this 3-valued logic. They are defined in the truth tables below:

P \ Q	T	F	⊥
		P∧Q	
T	T	F	⊥
F	F	F	F
⊥	⊥	F	⊥

Fig. 3.1. Conjunction

P \ Q	T	F	⊥
		P∨Q	
T	T	T	T
F	T	F	⊥
⊥	T	⊥	⊥

Fig. 3.2. Disjunction

   The conjunction of P and Q is true when both P and Q are true; false if one of P or Q is false, and undefined otherwise. The operation is commutative. The disjunction of P and Q (P ∨ Q) is true if one of P or Q is true; false if both P and Q are false; and undefined otherwise. The implication operation (P ⇒ Q) is true when P is false or when Q is true; it is undefined otherwise. The equivalence operation (P ≡ Q) is true when both P and Q are true or false; it is false when P is true and Q is false (or vice versa); it is undefined otherwise.

P \ Q	T	F	⊥
		P⇒Q	
T	T	F	⊥
F	T	T	T
⊥	T	⊥	⊥

Fig. 3.3. Implication

P \ Q	T	F	⊥
		P≡Q	
T	T	F	⊥
F	F	T	⊥
⊥	⊥	⊥	⊥

Fig. 3.4. Equivalence

The not operator (¬) is a unary operator such ¬A is true when A is false, false when A is true and undefined when A is undefined.

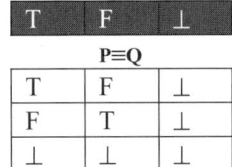

A	¬A
T	F
F	T
⊥	⊥

Table 3.8. Negation

---

[8] The above expression evaluates to true for Parnas logic (a 2-valued logic).

It is clear from the truth table definitions that the result of the operation may be known immediately after knowing the value of one of the operands (e.g., disjunction is true if P is true irrespective of the value of Q). The law of the excluded middle: i.e., $A \lor \neg A$ = true does not hold in the 3-valued logic of partial functions. However, this is reasonable [Jon:90] as one would not expect the following to be true:

$$(^1\!/_0 = 1) \lor (^1\!/_0 \neq 1).$$

There are other well-known laws that fail to hold such as:

(i) $E \Rightarrow E$

(ii)   Deduction theorem $E_1 \vdash E_2$ does not justify $\vdash E_1 \Rightarrow E_2$ unless it is known that E_1 is defined.

(iii)   Many of the tautologies of standard logic.

3.4.2   Parnas Logic

Predicate logic is an essential part of tabular expressions (discussed briefly in Chapter 2 and in detail in Chapter 8). It is employed to partition the domain in order to give the piecewise definition of a function. The Parnas approach to logic is based on the philosophy that truth-values should be true or false only.[9] That is, a logical term may be true or false only, and that there is no third logical value. The evaluation of a logical expression yields the value true or false irrespective of the assignment of values to the variables in the expression. The approach allows the following expression: $(y = \sqrt{x}) \lor (y = \sqrt{-x})$ that is undefined in classical logic to yield the value true.

A key advantage of the Parnas approach is that it does not introduce any new symbols into the logic (such as the undefined symbol $\bot$). Further, the logical connectives retain their traditional meaning. The logic should therefore be easy for engineers and computer scientists to understand as it is similar to the logic studied at high school.

The approach is to define the meaning of predicate expressions by first defining the meaning of the primitive predicate expressions. The primitive expressions are then used as the building bocks for predicate expressions. The evaluation of a primitive expression $R_j(V)$ (where V is a comma-separated set of terms with some elements of V including the application of partial functions) is false if the value of an argument of a function used in one of the terms of V is not in the domain of that function.[10] The following examples should make this clearer:

---

[9] However, I find it quite strange and unintuitive to assign the value false to the primitive predicate calculus expression $y = {}^1\!/_0$. I wonders where the classical engineering justification is for this approach is?

Expression	$x < 0$	$x \geq 0$
$y = \sqrt{x}$	False	True if $y = \sqrt{x}$, False otherwise
$y = {}^1/_0$	False	False
$y = x^2 + \sqrt{x}$	False	True if $y = x^2 + \sqrt{x}$, False otherwise

Table 3.9. Examples of Undefinedness

Expression	$i \in \{1 .. N\}$	$i \notin \{1..N\}$
$B[i] = x$	True if $B[i]=x$	False
$\exists i, B[i] = x$	True if $B[i]=x$ for some i, False otherwise	False

Table 3.10. Example of Undefinedness in Array

These primitive expressions are used to build the predicate expressions and the standard logical connectives are used to yield truth-values for the predicate expression.

The power of the Parnas logic is best demonstrated by considering a tabular expressions example taken from [Par:93]. The table below specifies the behavior of a program that searches the array B for the value x. The table describes the properties of the values of j' and *present'*. There are two cases to consider:

      1. There is an element in the array with the value of x;
      2. There is no such element in the array with the value of x.

	$(\exists\, i, B[i]=x)$	$\neg(\exists\, i,\ B[i]=x)$	
			H_2
H_1 $j'|$	$B[j']=x$	*true*	G
present'=	*true*	*false*	

Fig 3.5. Finding Index in Array

Clearly, from the example above the predicate expressions $\exists i, B[i] = x$ and $\neg(\exists\, i,\ B[i]=x)$ are defined. One disadvantage of the Parnas approach is that some common relational operators (e.g., $>, \geq, \leq$, and $<$) are not primitive in the logic. However, these relational operators are then constructed from primitive operators. Further, the axiom of reflection does not hold in the logic.

---

[10] The approach avoids the undefined logical value ($\perp$) and a 2-valued logic is maintained. The Parnas approach to undefined expressions seems to work well with his tabular expressions.

3.4.3   Dijkstra and Undefinedness

The **cand** and **cor** operators were introduced by Dijkstra to deal with undefined values. They are non-commutative operators and allow the evaluation of predicates that contain undefined values. For example, the expression x/y is undefined for $y = 0$. Consider the following expression:

$$y = 0 \vee (x/y = 2)$$

Then this expression is undefined when $y = 0$ as x/y is undefined, and the logical or operation is not defined when one of its operands is undefined. However, there is a case for giving meaning to such an expression when $y = 0$ since in that case the first operand of the logical or operation is true. Further, the logical or operation is defined to be true if either of its operands is true. This motivates the introduction of the **cand** and **cor** operators. These operators are associative and their truth tables are defined below:

a	b	a cand b
T	T	T
T	F	F
T	U	U
F	T	F
F	F	F
F	U	F
U	T	U
U	F	U
U	U	U

Table 3.11. *a cand b*

a	b	a cor b
T	T	T
T	F	T
T	U	T
F	T	T
F	F	F
F	U	U
U	T	U
U	F	U
U	U	U

Table 3.12. *a cor b*

The order of the evaluation of the operands for the **cand** operation is to evaluate the first operand; if the first operand is true then the result of the operation is the second operand; otherwise the result is false. The expression *a cand b* is equivalent to:

$$a \ \textbf{cand} \ b \ \cong \ \textbf{if} \ a \ \textbf{then} \ b \ \textbf{else} \ F$$

The order of the evaluation of the operands for the *cor* operation is to evaluate the first operand. If the first operand is true then the result of the operation is true; otherwise the result of the operation is the second operand. The expression *a* **cor** *b* is equivalent to:

$$a \ \textbf{cor} \ b \ \cong \ \textbf{if} \ a \ \textbf{then} \ T \ \textbf{else} \ b$$

The **cand** and **cor** operators satisfy the following laws:

- *Associativity*

    The **cand** and **cor** operators are associative.

    $$(A \text{ cand } B) \text{ cand } C = A \text{ cand } (B \text{ cand } C)$$
    $$(A \text{ cor } B) \text{ cor } C = A \text{ cor } (B \text{ cor } C)$$

- *Distributivity*

    The **cand** operator distributes over the **cor** operator and vice versa.

    $$A \text{ cand } (B \text{ cor } C) = (A \text{ cand } B) \text{ cor } (A \text{ cand } C)$$
    $$A \text{ cor } (B \wedge C) = (A \text{ cor } B) \text{ cand } (A \text{ cor } C)$$

De Morgan's law enables logical expressions to be simplified.

$$\neg (A \text{ cand } B) = \neg A \text{ cor } \neg B$$
$$\neg (A \text{ cor } B) = \neg A \text{ cand } \neg B$$

COMMENT (UNDEFINEDNESS)
It is best to avoid undefinedness and the preconditions of a function needs to be checked to ensure that the function is defined for the particular value. I consider it unintuitive to assign the truth value of false to the expression $y = {}^1\!/_0$ as proposed by Parnas. The approaches of Jones and Dijkstra have the disadvantage that they are 3-valued logic that are less intuitive than classical 2-valued logic.

3.5 Miscellaneous

There are other logics that arise in computer science. These include the temporal logics that are concerned with expressing properties that are time dependent. For example, a specification may require a fairness property to be expressed, and some formal methodists have introduced special temporal operators to express conditions that will always be true; conditions that will eventually be true; and a condition that will be true in the next time instance. For example:

$\Box$ P   - P is always true
$\Diamond$ P   - P is true sometime in the future
$\bigcirc$ P   - P is true in the next time instant.

It is also possible to express temporal operations directly in classical mathematics rather than applying specialist operators as above. This approach is preferred by Parnas and others. For example, the value of a function f at a time instance prior to the current time t is defined as:

$$Prior(f, t) = \lim_{t \to 0} f(t-\varepsilon)$$

Another logic that arises in computer science is fuzzy logic, and this logic is used to deal with degrees of truth. The reader is referred to texts on temporal logic and fuzzy logics.

Perhaps, one of the more unusual logics that has been invented is intuitionist logic.[11] The reader is referred to [Hey:66]. This logic has been applied to Type Theory by Martin Lof [Lof:84].

3.6 Tools for Logic

There are several tools available for theorem proving. These include the Boyer-Moore theorem prover known as NQTHM; the Isabelle theorem prover, and the HOL system. The Boyer-Moore theorem prover was developed in the early 1970s by B. S. Boyer and J. S. Moore. The tool has been continuously improved since then and it is currently known as NQTHM.[12] The theorem prover has been effective in proving well-known theorems such as Goedel's Incompleteness Theorem, the unsolvability of the Halting problem, a formalization of the Motorola MC 68020 Microprocessor, and many more. Computational Logic Inc.[13] is a company founded by Boyer and Moore in 1983 to share the benefits of a formal approach to software development with the wider computing community. It is based in Austin, Texas, and provides services in the mathematical modeling of hardware and software systems. This involves the use of mathematics and logic to formally specify microprocessors and other systems. The use of its theorem prover is then to formally verify the implementation meets its specification: i.e., to prove that the microprocessor or other system satisfies its specification.

Isabelle is a theorem proving environment developed at Cambridge University by Larry Paulson and Tobias Nipkow of the Technical University of

---

[11] Intuitionism was a highly controversial school of mathematics founded by the Dutch mathematician L. E. J. Brouwer. Initially, Brouwer was well-known for his work on topology including the fixpoint theorem. However, he was particularly interested in the foundations of mathematics and the resulting problems from the paradoxes of set theory. He was therefore interested in a secure foundations for mathematics and he took the view that for a mathematical object to be proved to exist then there must be a constructive way to produce the object. He therefore rejected the Law of the Excluded Middle and exhibited an extreme form of constructivism. This took the form of an absolute rejection of indirect proofs and argued that for an entity to exist then it needed to be constructed. Consequently, if the Brouwer view of the world were accepted then many of the classical theorems of mathematics (including his own well-known results in topology) could no longer be said to be true. He developed a form of logic called Intuitionist Logic in which many of the results of classical mathematics were no longer true.

[12] I understand that the NQTHM tool has been superseded by ACL2 available from the University of Texas.

[13] I understand that Computational Logic Inc. ceased trading in 1997.

Munich. It allows mathematical formulas to be expressed in a formal language and provides tools for proving those formulas. The main application is the formalization of mathematical proofs, and proving the correctness of computer hardware or software with respect to its specification and proving properties of computer languages and protocols. Isabelle is a generic theorem prover in the sense that it has the capacity to accept a variety of formal calculi, whereas most other theorem provers are specific to a specific formal calculus. Isabelle is available free of charge under an open source license.

The HOL system is an environment for interactive theorem proving in a higher-order logic.[14] The HOL system has been applied to the formalization of mathematics and the verification of hardware. It is used by academia and industry and is available free of change. HOL was originally developed at Cambridge University in the United Kingdom in the early 1980s, and it has been continuously improved upon since then. HOL 4 is the latest version and is an open source project.

There is a learning curve with the theorem-provers above and it generally takes a couple of months for users to become familiar with the theorem-prover.

3.7 Summary

This chapter considered propositional and predicate calculus. Propositional logic is the study of propositions, and a proposition is a statement that is either true or false. A formula in propositional calculus may contain several variables, and the truth or falsity of the individual variables and the meanings of the logical connectives determine the truth or falsity of the logical formula.

A rich set of connectives is employed in propositional calculus to combine propositions and to build the well-formed formulae of the calculus. This includes the conjunction of two propositions ($A \wedge B$), the disjunction of two propositions ($A \vee B$), and the implication of two propositions ($A \Rightarrow B$). These connectives allow compound propositions to be formed, and the truth of the compound propositions is determined from the truth-values of the constituent propositions and the rules associated with the logical connective. The meaning of the logical connectives is given by truth tables.

Predicates are statements involving variables, and these statements become propositions once these variables are assigned values. Predicate calculus

---

[14] Higher-order logic allows quantification over functions and predicates whereas quantification over variables only is allowed in first-order logic.

allows expressions, such as all members of the domain have a particular property to be expressed formally: e.g., $(\forall x)Px$, or that there is at least one member that has a particular property: e.g., $(\exists x)Px$. Predicate calculus may be employed to specify the requirements for a proposed system, to define a function that is piecewise defined and to prove the presence or absence of certain properties in a specification.

4
Z Specification Language

4.1 Introduction

Z is a formal specification language founded on Zermelo[1] set theory. It was developed at the Programming Research Group at Oxford University in the early 1980s [Dil:90]. It has been employed in both industry and academia and the language has been standardized in the ISO/IEC 13568:2000 standard. Z specifications are mathematical and it uses a classical 2-valued logic. This is a key advantage of its approach over the more conventional specification methods as Z specifications may be examined mathematically and various results about the specification proved. The use of mathematics helps to identify inconsistencies and gaps in the specification if they exist, and there are mathematical techniques to prove that the software implementation meets its specification.

Z is a model-oriented approach with an explicit model of the state of an abstract machine given, and operations are defined in terms of this state. Its main features include a mathematical notation that is used for the formal specification. The schema calculus is used to structure Z specifications and describes states and state transitions. It is visually striking, and consists essentially of boxes, with these boxes or schemas used to describe operations and states. The schema calculus enables schemas to be used as building blocks and combined with other schemas. The simple schema below is the specification of the positive square root of a real number.

$$
\begin{array}{|l}
-SqRoot\text{————} \\
\text{num?}, root! : \mathbb{R} \\
\hline
num? \geq 0 \\
root!^2 = num? \\
root! \geq 0 \\
\hline
\end{array}
$$

Fig. 4.1. Specification of Square Root

---

[1] Zermelo is well known for his work in axiomatic set theory. Fraenkel improved upon Zermelo's work and the resulting axiomatic system is known as Zermelo-Fraenkel (ZF) Set Theory. Zermelo is also well known for the proof of the well ordering theorem (i.e., every set can be well-ordered). He proved this result using the Axiom of Choice.

The schema calculus is a powerful means of decomposing a specification into smaller pieces or schemas. This decomposition helps to ensure that a Z specification is highly readable, as each individual schema is small in size and self-contained. Exception handling may be addressed by defining schemas for the exception cases, and then combining the exception schema with the original operation schema. Mathematical data types are used to model the data in a system, these data types obey mathematical laws. These laws enable simplification of expressions, and are useful with proofs.

Operations are defined in a precondition/postcondition style. A precondition must be true before the operation is executed and the postcondition must be true after the operation has been executed. In Z, the precondition is implicitly defined within the operation. The precondition for the specification of the square root function above is that $num? \geq 0$; i.e., the function *SqRoot* may be applied to positive real numbers only. Each operation has an associated proof obligation to ensure that if the precondition is true, then the operation preserves the system invariant. The system invariant is a property of the system that must be true at all times. The initial state itself is, of course, required to satisfy the system invariant. The postcondition for the square root function is $root!^2 = num?$ and $root! \geq 0$. That is, the square root of a number is positive and its square gives the number. Postconditions employ a logical predicate which relates the prestate to the poststate, and the poststate of a variable being distinguished by priming the variable, e.g., v'.

Z is a typed language and whenever a variable is introduced its type must be given. A type is simply a collection of objects, and there are several standard types in Z. These include the natural numbers $\mathbb{N}$, the integers $\mathbb{Z}$, and the real numbers $\mathbb{R}$. The declaration of a variable x of type X is written $x : X$. It is also possible to create your own types in Z. Various conventions are employed within Z specification, for example $v?$ indicates that v is an input variable; $v!$ indicates that v is an output variable. The variable $num?$ is an input variable and $root!$ is an output variable for the square root example above. The notation Ξ in a schema indicates that the operation *Op* does not affect the state; whereas the notation Δ in the schema indicates that *Op* is an operation that affects the state.

Many of the data types employed in Z have no counterpart in standard programming languages. It is therefore important to identify and describe the concrete data structures that ultimately will represent the abstract mathematical structures. As the concrete structures may differ from the abstract, the operations on the abstract data structures may need to be refined to yield operations on the concrete data that yield equivalent results. For simple systems, direct refinement (i.e., one step from abstract specification to implementation) may be possible; in more complex systems, deferred refinement[2] is employed, where a sequence of increasingly concrete specifications are produced to yield the executable specification.

---

[2] It is debatable whether refinement is cost effective in mainstream software engineering. It involves producing a sequence of increasingly more concrete specifications until eventually the executable code is produced. Each refinement step has associated proof obligations to prove that the refinement step is valid.

EXAMPLE 4.1
The following is a *Z* specification to borrow a book from a library system. The library is made up of books that are on the shelf, books that are borrowed, and books that are missing. These are three mutually disjoint subsets of the set of books *Bkd-Id*.

   The system state is defined in the *Library* schema below, and operations such as *Borrow* and *Return* affect the state. The *Borrow* operation is specified below. There is a calculus for combining schemas to make larger specifications, and this is discussed later in the chapter.

$$
\begin{array}{|l}
\textit{–Library}\text{————} \\
\textit{on-shelf, missing, borrowed} : \mathbb{P}\ \textit{Bkd-Id} \\
\hline
\textit{on-shelf} \cap \textit{missing} = \emptyset \\
\textit{on-shelf} \cap \textit{borrowed} = \emptyset \\
\textit{borrowed} \cap \textit{missing} = \emptyset \\
\hline
\end{array}
$$

Fig. 4.2. Specification of a Library System

The notation $\mathbb{P}\ \textit{Bkd-Id}$ is used to represent the power set of *Bkd-Id* (i.e., the set of all subsets of *Bkd-Id*). The disjointness condition for the library is expressed by the requirement that the pairwise intersection of the subsets *on-shelf, borrowed, missing* is the empty set.

$$
\begin{array}{|l}
\textit{–Borrow}\text{————} \\
\Delta\ \textit{Library} \\
\textit{b?} : \textit{Bkd-Id} \\
\hline
\textit{b?} \in \textit{on-shelf} \\
\textit{on-shelf'} = \textit{on-shelf} \setminus \{\textit{b?}\} \\
\textit{borrowed'} = \textit{borrowed} \cup \{\textit{b?}\} \\
\hline
\end{array}
$$

Fig. 4.3. Specification of Borrow Operation

   The specification of the library system models a library with sets representing books on the shelf, on loan, or missing. The precondition for the *Borrow* operation is that the book must be available on the shelf to borrow. The postcondition is that the borrowed book is added to the set of borrowed books and is removed from the books on the shelf.

   Z has been successfully applied in industry, and one of its well-known successes is the CICS project at IBM Hursley in the United Kingdom.[3]

---

[3] This project claimed a 9% increase in productivity attributed to the use of formal methods.

4.2 Sets

Sets have been discussed earlier in the book and this section focuses on their use in Z. Sets may be enumerated by listing all of their elements. Thus, the set of all even natural numbers less than or equal to 10 is:

$$\{2,4,6,8,10\}.$$

Sets can be created from other sets using set comprehension. For example, the set of even natural numbers less than 10 is given by set comprehension as:

$$\{n : \mathbb{N} \mid n \neq 0 \wedge n < 10 \wedge n \bmod 2 = 0 \bullet n\}$$

There are three main parts to the set comprehension above. The first part is the signature of the set and this is given by $n : \mathbb{N}$ above. The first part is separated from the second part by a vertical line. The second part is given by a predicate and this is $n \neq 0 \wedge n < 10 \wedge n \bmod 2 = 0$ in the example. The second part is separated from the third part by a bullet. The third part is a term, and is simply n in the example above. The term could be a more complex expression: e.g., $\log(n^2)$.

In mathematics, there is just one empty set. However, since Z is a typed set theory, there is an empty set for each type of set. Hence, there are an infinite number of empty sets, one for each type of set. The empty set is written $\varnothing$ [X] where X is the type of the empty set. In practice, X is omitted when the type is clear.

Various operations on sets such as union, intersection, difference, and symmetric difference are employed in Z. These operations have been discussed earlier in Chapter 2. The powerset of a set X is the set of all subsets of X. It is denoted by $\mathbb{P}$ X and includes the empty set. The set of nonempty subsets of X is denoted by $\mathbb{P}_1$ X where

$$\mathbb{P}_1 \, X \; == \; \{U : \mathbb{P} \, X \mid U \neq \varnothing \, [X]\}.$$

A finite set of elements of type X (denoted by *F* X) is a subset of X that cannot be put into one-to-one correspondence with a proper subset of itself. This is defined formally as:

$$\boldsymbol{F} \, X \; == \; \{U : \mathbb{P} \, X \mid \neg \exists V : \mathbb{P} \, U \bullet V \neq U \wedge (\exists f : V \rightarrowtail\!\!\!\!\rightarrow U)\}.$$

The expression $f : V \rightarrowtail\!\!\!\!\rightarrow U$ denotes that f is a bijection from U to V and injective, surjective, and bijective functions are discussed later.

The fact that Z is a typed language means that whenever a variable is introduced (e.g., in quantification with $\forall$ and $\exists$) it is first declared. For example, $\forall$

j:J • P $\Rightarrow$Q. There is also the unique existential quantifier $\exists_1$ *j*:J | P • Q which states that there is exactly one *j* of type J that has property P.

4.3 Relations

Relations have been discussed earlier in the book and they are used extensively in *Z*. A relation R between X and Y is any subset of the Cartesian product of X and Y; i.e., R $\subseteq$ (X $\times$ Y) and the relation is denoted by R : X $\leftrightarrow$Y. The notation $x \mapsto y$ indicates that the pair $(x,y) \in$ R.

Consider, the relation *home_owner* : *Person* $\leftrightarrow$ *Home* that exists between people and their homes. An entry *daphne* $\mapsto$ *mandalay* $\in$ *home_owner* if *daphne* is the owner of *mandalay*. It is possible for a person to own more than one home:

rebecca $\mapsto$ *nirvana* $\in$ *home_owner*

rebecca $\mapsto$ *tivoli* $\in$ *home_owner*

It is possible for two people to share ownership of a home:

sheila $\mapsto$ *nirvana* $\in$ *home_owner*

blaithín $\mapsto$ *nirvana* $\in$ *home_owner*

There may be some people who do not own a home and there is no entry for these people in the relation *home_owner*. The type *Person* includes every possible person, and the type *Home* includes every possible home. The domain of the relation *home_owner* is given by:

$x \in$ dom *home_owner* $\Leftrightarrow \exists h$: *Home* • $x \mapsto h \in$ *home_owner*.

The range of the relation *home_owner* is given by:

$h \in$ ran *home_owner* $\Leftrightarrow \exists x$: *Person* • $x \mapsto h \in$ *home_owner*.

The composition of two relations *home_owner* : *Person* $\leftrightarrow$ *Home* and *home_value* : *Home* $\leftrightarrow$ *Value* yields the relation *owner_wealth* : *Person* $\leftrightarrow$ *Value* and is given by the relational composition *home_owner* ; *home_value* where:

$p \mapsto v \in$ *home_owner* ; *home_value* $\Leftrightarrow$

($\exists h$: *Home* • $p \mapsto h \in$ *home_owner* $\wedge h \mapsto v \in$ *home_value*)

The relational composition may also be expressed as:

$$owner\_wealth = home\_value \circ home\_owner.$$

The union of two relations arises frequently in practice, for example, in the derivation of the after state of a relation from the before state. Suppose a new entry *aisling ↦ muckross* is to be added. Then this is given by

$$home\_owner' = home\_owner \cup \{aisling \mapsto muckross\}$$

Suppose that we are interested in knowing all females who are house owners. Then we restrict the relation *home_owner* so that the first element of all ordered pairs have to be female. Consider the sets *male, female* : $\mathbb{P}$ *Person* with $\{aisling, eithne\} \subseteq female,$ and *lawrence* $\subseteq$ *male,* and *male* $\cap$ *female* $= \emptyset$.

$$home\_owner = \{aisling \mapsto muckross, eithne \mapsto parknasilla,$$
$$lawrence \mapsto nirvana\}$$

$$female \lhd home\_owner = \{aisling \mapsto muckross, eithne \mapsto parknasilla\}$$

That is, *female* $\lhd$ *home_owner* is a relation that is a subset of *home_owner* and the first element of each ordered pair in the relation is female. The operation $\lhd$ is termed domain restriction and its fundamental property is:

$$x \mapsto y \in U \lhd R \Leftrightarrow (x \in U \wedge x \mapsto y \in R\}$$

where R : X $\leftrightarrow$ Y and $U : \mathbb{P}$ X.

There is also a domain antirestriction (subtraction) operation and its fundamental property is:

$$x \mapsto y \in U \ntriangleleft R \Leftrightarrow (x \notin U \wedge x \mapsto y \in R\}$$

where R : X $\leftrightarrow$ Y and $U : \mathbb{P}X$.

There are also range restriction (the $\rhd$ operator) and the range antirestriction operator (the $\ntriangleright$ operator). These are discussed in [Dil:90].

4.4 Functions

A function [Dil:90] is an association between objects of some type X and objects of another type Y such that given an object of type X, there exists only one object in Y associated with that object. A function is a set of ordered pairs where the first element of the ordered pair has at most one element associated with it. A function is therefore a special type of relation, and a function may be total or partial. A total function has exactly one element in Y associated with each element of X, whereas a partial function has at most one element of Y associated with each element of X (there may be elements of X that have no element of Y associated with them).

A partial function from X to Y (denoted $f : X \nrightarrow Y$) is a relation $f : X \leftrightarrow Y$ such that:

$$\forall x{:}X; \, y,z{:}Y \bullet (x \mapsto y \in f \wedge x \mapsto z \in f \Rightarrow y = z).$$

The association between x and y is denoted by $f(x) = y$, and this indicates that the value of the partial function f at x is y. A total function from X to Y (denoted $f : X \rightarrow Y$) is a partial function such that every element in X is associated with some value of Y.

$$f : X \rightarrow Y \Leftrightarrow f : X \nrightarrow Y \wedge \operatorname{dom} f = X$$

Clearly, every total function is a partial function.

One operation that arises quite frequently in specifications is the function override operation. Consider the following specification of a temperature map:

$$
\begin{array}{|l}
\hline
\text{—\,TempMap—————} \\
CityList : \mathbb{P} \;\; City \\
temp : City \nrightarrow Z \\
\hline
\operatorname{dom} temp = CityList \\
\hline
\end{array}
$$

Fig. 4.4. Temperature Map

Suppose the temperature map is given by $temp = \{Cork \mapsto 17,\ Dublin \mapsto 19,\ Mallow \mapsto 15\}$. Then consider the problem of updating the temperature map if a new temperature reading is made in Cork say $\{Cork \mapsto 18\}$. Then the new temperature chart is obtained from the old temperature chart by function override to yield $\{Cork \mapsto 18,\ Dublin \mapsto 19,\ Mallow \mapsto 15\}$. This is written as:

$$temp' = temp \oplus \{Cork \mapsto 18\}.$$

The function override operation combines two functions of the same type to give a new function of the same type. The effect of the override operation is that the entry $\{Cork \mapsto 17\}$ is removed from the temperature chart and replaced with the entry $\{Cork \mapsto 18\}$.

Suppose $f,g : X \nrightarrow Y$ are partial functions then $f \oplus g$ is defined and indicates that f is overridden by g. It is defined as follows:

$$(f \oplus g)\ (x) = g(x) \quad \text{where } x \in \text{dom } g$$
$$(f \oplus g)\ (x) = f(x) \quad \text{where } x \notin \text{dom } g \wedge x \in \text{dom } f.$$

This may also be expressed (using function override) as:

$$f \oplus g = ((\text{dom } g) \ntriangleleft f) \cup g.$$

There is notation in Z for injective, surjective, and bijective functions. An injective function is one to one: i.e.,

$$f(x) = f(y) \Rightarrow x = y.$$

A surjective function is onto: i.e.,

$$\text{Given } y \in Y, \exists x \in X \text{ such that } f(x) = y.$$

A bijective function is one to one and onto and indicates that the sets X and Y can be put into one to one correspondence with one another. Z includes λ-notation to define functions. For example, cube $= \lambda x{:}\mathbf{N} \bullet x * x \ * x$. Function composition $f\,;\,g$ is similar to relational composition.

4.5 Sequences

The type of all sequences of elements drawn from a set X is denoted by seq X. Sequences are written as $\langle x_1, x_2, \ .... \ x_n \rangle$ and the empty sequence is denoted by $\langle \rangle$. Sequences may be used to specify the changing state of a variable over time where each element of the sequence represents the value of the variable at a discrete time instance.

Sequences are functions and a sequence of elements drawn from a set X is a finite function from the natural numbers to X. A partial finite function f from X to Y is denoted by $f : X \nrightarrow Y$. A finite sequence of elements of X is given by $f : \mathbf{N} \nrightarrow X$, and the domain of the function consists of all numbers between 1 and $\#f$. It is defined formally as:

$$\text{seq } X == \{f : \mathbf{N} \nrightarrow X \mid \text{dom } f = 1 \ .. \ \#f \bullet f\}.$$

The sequence $\langle x_1, x_2,, ..., x_n \rangle$ above is given by:

$$\{1 \mapsto x_1, 2 \mapsto x_2, ..., n \mapsto x_n\}$$

There are various functions to manipulate sequences. These include the sequence concatenation operation. Suppose $\sigma = \langle x_1, x_2, ..., x_n \rangle$ and $\tau = \langle y_1, y_2, ..., y_m \rangle$ then:

$$\sigma \frown \tau = \langle x_1, x_2, ..., x_n, y_1, y_2, ..., y_m \rangle$$

The head of a nonempty sequence gives the first element of the sequence.

$$\text{head } \sigma = \text{head } \langle x_1, x_2, ..., x_n \rangle = x_1.$$

The tail of a non-empty sequence is the same sequence except that the first element of the sequence is removed.

$$\text{tail } \sigma = \text{tail } \langle x_1, x_2, ..., x_n \rangle = \langle x_2, ..., x_n \rangle$$

Suppose $f : X \rightarrow Y$ and σ is a sequence (i.e., $\sigma : \text{seq } X$) then the function map applies f to each element of σ:

$$\text{map } f \ \sigma = \text{map } f \ \langle x_1, x_2, ..., x_n \rangle = \langle f(x_1), f(x_2), ..., f(x_n) \rangle$$

The map function may also be expressed via function composition as:

$$\text{map } f \ \sigma = \sigma \, ; f.$$

The reverse order of a sequence is given by the rev function:

$$\text{rev } \sigma = \text{rev } \langle x_1, x_2, ..., x_n \rangle = \langle x_n, ..., x_2, x_1 \rangle.$$

4.6 Bags

A bag is similar to a set except that there may be multiple occurrences of each element in the bag. A bag of elements of type X is defined as a partial function from the type of the elements of the bag to positive whole numbers. The definition of a bag of type X is:

$$\text{bag } X == X \nrightarrow \mathbb{N}_1.$$

For example, a bag of marbles may contain 3 blue marbles, 2 red marbles, and 1 green marble. This is denoted by $B = [\![\,b,b,b,g,r,r\,]\!]$. The bag of marbles is thus denoted by:

$$bag\ Marble == Marble \mapsto \mathbb{N}_1.$$

The function count determines the number of occurrences of an element in a bag, for the example above, count *Marble* $b = 3$, and count *Marble* $y = 0$ since there are no yellow marbles in the bag. This is defined formally as:

$$count\ bag\ X\ \ y = 0 \qquad\qquad y \notin bag\ X$$
$$count\ bag\ X\ \ y = (bag\ X)\,(y) \qquad\qquad y \in bag\ X.$$

An element y is in bag X if and only if y is in the domain of bag X.

$$y\ in\ bag\ X \Leftrightarrow y \in dom\ (bag\ X).$$

The union of two bags of marbles $B_1 = [\![\,b,b,b,g,,r,r\,]\!]$ and $B_2 = [\![\,b,,g,,r,y\,]\!]$ is given by $B_1 \uplus B_2 = [\![\,b,b,b,b,g,g,r,r,r,y\,]\!]$. It is defined formally as:

$$(B_1 \uplus B_2)\,(y) = \quad B_2\,(y) \qquad\qquad y \notin dom\ B_1 \wedge y \in dom\ B_2$$
$$(B_1 \uplus B_2)\,(y) = \quad B_1\,(y) \qquad\qquad y \in dom\ B_1 \wedge y \notin dom\ B_2$$
$$(B_1 \uplus B_2)\,(y) = \quad B_1\,(y) + B_2\,(y) \qquad y \in dom\ B_1 \wedge y \in dom\ B_2.$$

A bag may be used to record the number of occurrences of each product in a warehouse as part of an inventory system. The number of items remaining for each product in a vending machine may be modeled by a bag.

$$
\begin{array}{|l}
\hline
-\Delta Vending\ Machine\text{------} \\
stock : bag\ Good \\
price : Good \rightarrow \mathbb{N}_1 \\
\hline
dom\ stock \subseteq dom\ price \\
\hline
\end{array}
$$

Fig. 4.5. Specification of Vending Machine Using Bags

The operation of a vending machine would require other operations such as identifying the set of acceptable coins, checking that the customer has entered sufficient coins to cover the cost of the good, returning change to the customer, and updating the quantity on hand of each good after a purchase. A more detailed examination is in [Dil:90].

4.7 Schemas and Schema Composition

Schema are used for specifying states and state transitions and they group all relevant information that belongs to a state description. They employ notation to represent before and after state (e.g., s and s'). The schemas in Z are visually striking and the specification is presented in 2-dimensional graphic boxes.

There are a number of schema operations and conventions that allow the specification of complex operations concisely and help to make Z specifications readable. These include schema inclusion, linking schemas with propositional connectives and the Δ and Ξ conventions. Schema composition is analogous to relational composition, and allows new schemas to be derived from existing schemas.

A schema name S_1 may be included in the declaration part of another schema S_2. The effect of the inclusion is that the declarations in S_1 are now part of S_2 and the predicates of S_1 are S_2 are joined together by conjunction. If the same variable is defined in both S_1 and S_2, then it must be of the same type in both schemas.

$$
\begin{array}{|l}
\hline
S_1 \\
\hline
x,y : \mathbb{N} \\
\hline
x + y > 2 \\
\hline
\end{array}
\qquad
\begin{array}{|l}
\hline
S_2 \\
\hline
S_1;\, z : \mathbb{N} \\
\hline
z = x + y \\
\hline
\end{array}
$$

Fig. 4.6. Specification of S_1 and S_2

The result is that S_2 includes the declarations and predicates of S_1.

$$
\begin{array}{|l}
\hline
S_2 \\
\hline
x,y : \mathbb{N} \\
z : \mathbb{N} \\
\hline
x + y > 2 \\
z = x + y \\
\hline
\end{array}
$$

Fig. 4.7. Schema Inclusion

Two schemas may be linked by propositional connectives such as $S_1 \wedge S_2$, $S_1 \vee S_2$, $S_1 \Rightarrow S_2$, and $S_1 \Leftrightarrow S_2$. The schema $S_1 \vee S_2$ is formed by merging the declaration of S_1 and S_2, and then combining their predicates by the logical $\vee$ operator. For example, $S = S_1 \vee S_2$ yields:

$$
\begin{array}{|l}
\hline
S \\
\hline
x,y : \mathbb{N} \\
z : \mathbb{N} \\
\hline
x + y > 2 \vee z = x + y \\
\hline
\end{array}
$$

Fig. 4.8. Merging Schemas ($S_1 \vee S_2$)

The schema inclusion and linking of schemas employ normalization to convert subtypes to maximal types and to employ a predicate to restrict the maximal type to the subtype. This involves replacing declarations of variables (e.g., $u : 1$ ..35 with $u : Z$ and adding the predicate $u > 0$ and $u < 36$ to the predicate part of the schema). A more detailed explanation is in [Dil:90].

The Δ and Ξ conventions are used extensively in schemas. The notation Δ *TempMap* is used in the specification of schemas that involve a change of state. It represents:

$$\Delta\, TempMap = TempMap \wedge TempMap\,'$$

$$
\begin{array}{|l}
-\Delta TempMap\text{\rule{1.5cm}{0.4pt}} \\
CityList, CityList' : \mathbb{P}\; City \\
temp, temp' : City \mapsto Z \\
\hline
\mathrm{dom}\; temp = CityList \\
\mathrm{dom}\; temp' = CityList' \\
\hline
\end{array}
$$

Fig. 4.9. Specification of $\Delta TempMap$ Schema

The notation Ξ *TempMap* is used in the specification of operations that do not involve a change to the state. It represents:

$$
\begin{array}{|l}
-\Xi\, TempMap\text{\rule{1.5cm}{0.4pt}} \\
\Delta TempMap \\
\hline
CityList = CityList' \\
temp = temp' \\
\hline
\end{array}
$$

Fig. 4.10. Specification of $\Xi\, TempMap$ Schemas

Schema composition is analogous to relational composition and allows new specifications to be built from existing ones. It allows a way of relating the after state variables of one schema with the before variables of another schema. The composition of two schemas S and T (S ; T) is described in detail in [Dil:90] and involves four steps:

Step	Procedure
1.	Rename all *after* state variables in S to something new: $S\,[s^+/s\,']$.
2.	Rename all *before* state variables in T to the same new thing: i.e., $T\,[s^+/s]$.

3.	Form the conjunction of the two new schemas: $S\ [s^+/s\ '] \wedge T\ [s^+/s].$
4.	Hide the variable introduced in step 1 and 2. $S\ ;\ T = (S\ [s^+/s\ '] \wedge T\ [s^+/s]) \setminus (s^+)$

Table 4.1. Schema Composition

The example below is adapted from [Dil:90] and should make schema composition clearer. Consider the composition of S and T where S and T are defined as follows:

-S——
$x, x', y?\ :\ \mathbb{N}$
———
$x' = y? - 2$

-T——
$x, x'\ :\ \mathbb{N}$
———
$x' = x + 1$

-S_1——
$x, x^+, y?\ :\ \mathbb{N}$
———
$x^+ = y? - 2$

-T_1——
$x^+, x'\ :\ \mathbb{N}$
———
$x' = x^+ + 1$

Fig. 4.11. Specification of S_1 and T_1

S_1 and T_1 represent the results of step 1 and step 2 in Table 4.1 above. It involves renaming x' to x^+ in S, and then renaming x to x^+ in T. Step 3 and step 4 of Table 4.1 yield:

-$S_1 \wedge T_1$——
$x, x^+, x', y?\ :\ \mathbb{N}$
———
$x^+ = y? - 2$
$x' = x^+ + 1$

-$S\ ;\ T$——
$x,\ x',\ y?\ :\ \mathbb{N}$
———
$\exists x^+ : \mathbb{N} \bullet$
$(x^+ = y? - 2$
$x' = x^+ + 1)$

Fig. 4.12. Schema Composition

Schema composition is useful as it allows new specifications to be created from existing ones.

4.8 Reification and Decomposition

A Z specification involves defining the state of the system and then specifying various operations. The formal specification is implemented by a programmer, and mathematical proof may be employed to prove that a program meets its specification. The Z specification language employs many constructs that are not part of conventional programming languages, and operations are specified by giving their preconditions and postconditions.

Hence, there is a need to write an intermediate specification that is between the original Z specification and the eventual program code. This intermediate specification is more algorithmic and uses less abstract data types than a Z specification. The intermediate specification is termed the design and the design needs to be correct with respect to the specification, and the program needs to be correct with respect to the design. The design is a refinement (reification) of the state of the specification, and the operations of the specification have been decomposed into those of the design.

The representation of an abstract data type like a set by a sequence is termed data reification, and data reification is concerned with the process of transforming an abstract data type into a concrete data type. The abstract and concrete data types are related by the retrieve function, and the retrieve function maps the concrete data type to the abstract data type. There are typically several possible concrete data types for a particular abstract data type (i.e., refinement is a relation), whereas there is one abstract data type for a concrete data type (i.e., retrieval is a function). For example, sets are often reified to unique sequences; however, more than one unique sequence can represent a set, whereas a unique sequence represents exactly one set.

The operations defined on the concrete data type need to be related to the operations defined on the abstract data type. The commuting diagram property is required to hold; i.e., for an operation $\boxdot$ on the concrete data type to correctly model the operation $\odot$ on the abstract data type it is required that the following property holds:

$$ret\ (\sigma \boxdot \tau) = (ret\ \sigma) \odot (ret\ \tau).$$

This is known as the commuting diagram property and it requires a proof that the diagram in Fig. 4.13 below commutes.

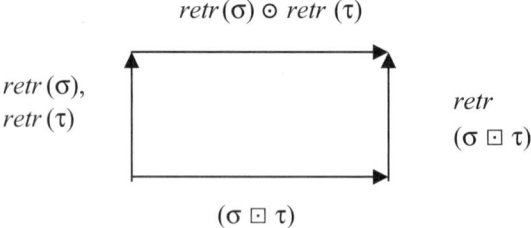

$$retr\,(\sigma) \odot retr\ (\tau)$$

$$retr\,(\sigma),$$
$$retr\,(\tau)$$

$$retr$$
$$(\sigma \boxdot \tau)$$

$$(\sigma \boxdot \tau)$$

Fig. 4.13. Refinement Commuting Diagram

In Z, the refinement and decomposition is done with schemas. It is required to prove that the concrete schemas are a valid refinement of the abstract schemas, and this gives rise to a number of proof obligations. It needs to be proved that the initial states correspond to one another, and that each operation in the concrete schema is correct with respect to the operation in the abstract schema, and also that it is applicable (i.e., whenever the abstract operation may be performed the concrete operation may be performed also).

4.9 Proof in *Z*

Mathematicians employ rigorous proofs of theorems using technical and natural language. Logicians employ formal proofs to prove theorems using propositional and predicate calculus. Formal proofs generally involve a long chain of reasoning with every step of the proof justified. Long formal proofs require tool support to prevent errors, as it is easy to make an invalid assumption in deduction. Rigorous proofs [Dil:90] have been described as being analogous to high level programming languages and involve precise reasoning, whereas formal proofs are analogous to machine language.

   A rigorous mathematical proof includes natural language and mathematical symbols and often many of the tedious details of the proof are omitted. Most proofs in formal methods such as *Z* are concerned with cross-checking on the details of the specification or the validity of the refinement, or proofs that certain properties are satisfied by the specification. There are many tedious lemmas to be proved, and tool support is therefore essential, as proofs by hand are notorious for containing errors or jumps in reasoning. Machine proofs are lengthy and unreadable, but help to avoid errors as every step in the proof needs to be justified.

   A formal mathematical proof consists of a sequence of formulae, where each element is either an axiom or derived from a previous element in the series by applying a fixed set of mechanical rules. The proof of various properties about the programs increases confidence in the correctness of the program. Examples of proofs required during software development with *Z* include:

- Proof of invariant preservation
- Proof of validity of refinement steps
- Proof that specific properties are preserved in the refinement.

Tool support is required for software development with Z and this is discussed in the next section.

4.10 Tools for *Z*

There are a number of tools that support software development with *Z*. Perhaps, the most widely known is the *Z*/EVES tool that is available from ORA[4] in Canada. The main features of the *Z*/EVES tool are:

Feature	Description
Syntax checker	Checks whether the *Z* syntax is correct.
Type checker	Performs type checking.
Schema expansion	*Z*/EVES allows a schema definition to be expanded and simplified. The references in the schema are expanded and the theorem prover is employed to simplify the results of the expansion.
Precondition calculation	The precondition is implicit in a *Z* specification and *Z*/EVES allows the precondition of a schema to be calculated. This requires interaction with the theorem prover.
Domain checking	These are proof obligations to ensure that the operation is defined on its domain. The objective is to show that the expression is defined on its domain and is therefore meaningful. This helps to avoid problems with undefinedness.
Theorem prover	The *Z*/EVES theorem prover provides automated support for proving theorems as well as allowing the user to guide the theorem prover.

Table 4.2. Features of *Z*/EVES Tool

The tool includes a graphical user interface that allows the *Z* specifications to be entered. It supports most of the *Z* notation with the exception of the unique existence quantifier. The *Z*/EVES tool may be introduced gradually in an organization rather than using the full functionality of the tool initially.

The Fuzz tool is an academic tool produced by Mike Spivey and allows printing and type-checking of *Z* specifications. It is used with LaTeX and the *Z* specification is entered in an ASCII file (in LaTeX format). It is then checked for conformance to the *Z* standard and error messages are produced. It is available (at the cost of a license fee) from Mike Spivey at Oxford.

There are many other academic and semicommercial tools available for *Z* including RoZ that produces *Z* specifications from UML diagrams;[5] CADiZ

---

[4] ORA has decided to cease its distribution of Z/EVES from mid-2005. I am unclear at this stage as to how future distribution and support of Z/EVES will be handled. The tool has been distributed to users in over sixty two countries and it is available on Linux and Microsoft NT/XP.

that prepares *Z* specification and performs type checking and is available from the University of York;[6] HOL which is a mechanical theorem proving system based on higher-order logic and provides some support for *Z*; and ProofPower[7] which is a suite of tools supporting specification and proof in higher-order logic and *Z*. There is also an open-source project (CZT)[8] from the *Z* community that is aiming to build a set of tools for editing, typechecking, and animating formal specifications in the *Z* specification language.

4.11 Summary

Z was developed at the Programming Research Group at Oxford University and has been employed in both industry and academia. Z specifications are mathematical and use a classical 2-valued logic. The specifications may be examined mathematically and various results about the specification proved. Mathematics helps to identify inconsistencies and gaps in the specification if they exist.

Z is a model-oriented approach and an explicit model of the state of an abstract machine is given. The operations are defined in terms of their effect on the state. Its main features include the visually striking schema calculus and a mathematical notation that is similar to VDM. The schema calculus consists essentially of boxes, and these boxes or schemas are used to describe operations and states. The schema calculus enables schemas to be used as building blocks to form larger specifications.

The schema calculus is a powerful means of decomposing a specification into smaller pieces or schemas. This decomposition helps to ensure that a *Z* specification is highly readable, as each individual schema is small in size and self-contained. Exception handling may be addressed by defining schemas for the exception cases, and then combining the exception schema with the original operation schema. Mathematical data types are used to model the data in a system.

Z is a highly expressive specification language and includes notation for sets, functions, relations, bags, sequences, predicate calculus, and schema calculus. It has been employed successfully in academia and industry.

---

[5] This is a prototype tool produced as part of the Champollion project. I am not aware of any plan to commercialize the tool. I am unclear as to which UML diagrams the tool can handle. Further information is available on http://www-lsr.imag.fr/Les.Groupes/pfl/RoZ/.

[6] Information on CADiZ is available on http://www-users.cs.york.ac.uk/~ian/cadiz/.

[7] For further information on ProofPower please go to http://www.lemmaone.com/ProofPower/index/.

[8] Further information on CZT is available in http://czt.sourceforge.net/.

5
Vienna Development Method

5.1 Introduction

VDM dates from work done by the IBM research laboratory in Vienna in the 1960s. Their aim was to specify the semantics of the PL/1 programming language. This was achieved by employing the Vienna Definition Language (VDL), taking an operational semantic approach; i.e. the semantics of a language are determined in terms of a hypothetical machine which interprets the programs of that language [BjJ:82]. Later work led to the Vienna Development Method (VDM) with its specification language, Meta IV.[1] This concerned itself with the denotational semantics of programming languages; i.e., a mathematical object (set, function, etc.) is associated with each phrase of the language [BjJ:82]. The mathematical object is the *denotation* of the phrase. The initial application of VDM was to programming language semantics. Today, VDM is mainly employed to formally specify software and includes a development method.

    The Vienna group was broken up in the mid-1970s and this led to the formation of different schools of the VDM in diverse locations. These include the Danish school led by Dines Bjørner;[2] the English school led by Cliff Jones;[3] and the Polish school led by Andrez Blikle. The various schools of VDM are described in [Mac:90]. Further work on VDM and Meta-IV continued in the 1980s and an ISO standard (International Standard ISO/IEC 13817-1) for VDM appeared in December 1996.

    VDM is a m*odel-oriented approach* and this means that an explicit model of the state of an abstract machine is given, and operations are defined in terms of this state. Operations may act on the system state, taking inputs and producing outputs and a new system state. Operations are defined in a precondi-

---

[1] The name chosen is a pun on metaphor.

[2] Dines Bjørner's background is both academic (Technical University of Denmark abd Macau) and industrial (IBM). He was key note speaker at IWFM'98 held in Cork, Ireland.

[3] Cliff Jones has a background in academia (Manchester and Newcastle) and industrial (IBM and Harlequin). He was one of the key note speakers for the first IWFM (Irish Workshop in Formal Methods) series held in Dublin in 1997 for which the author was programme chair.

tion and postcondition style. Each operation has an associated proof obligation to ensure that if the precondition is true, the operation preserves the system invariant. The initial state itself is, of course, required to satisfy the system invariant. VDM uses keywords to distinguish different parts of the specification, e.g., preconditions and postconditions are introduced by the keywords *pre* and *post* respectively. In keeping with the philosophy that formal methods specifies *what* a system does as distinct from *how*, VDM employs postconditions to stipulate the effect of the operation on the state. The previous state is then distinguished by employing *hooked variables*, e.g., $\overset{\smile}{v}$, and the postcondition specifies the new state *(defined by a logical predicate relating the prestate to the poststate)*.

VDM is more than its specification language Meta IV (called VDM-SL in the standardization of VDM) and is, in fact, a development method, with rules to verify the steps of development. The rules enable the executable specification, i.e., the detailed code, to be obtained from the initial specification via refinement steps. Thus, we have a sequence $S = S_0, S_1, ..., S_n = E$ of specifications, where S is the initial specification, and E is the final (executable) specification:

$$S = S_0 \sqsubseteq S_1 \sqsubseteq S_2 \sqsubseteq \ ...\sqsubseteq\ S_n = E.$$

Retrieval functions enable a return from a more concrete specification, to the more abstract specification. The initial specification consists of an initial state, a system state, and a set of operations. The system state is a particular domain, where a domain is built out of primitive domains such as the set of natural numbers, etc., or constructed from primitive domains using domain constructors such as Cartesian product, disjoint union, etc. A domain-invariant predicate may further constrain the domain, and a *type* in VDM reflects a domain obtained in this way. Thus, a type in VDM is more specific than the signature of the type, and represents values in the domain defined by the signature, which satisfy the domain invariant. In view of this approach to types, it is clear that VDM types may not be "statically type checked".

VDM specifications are structured into modules, with a module containing the module name, parameters, types, operations, etc. Partial functions arise naturally in computer science. The problem is that many functions, especially recursively defined functions, can be undefined or fail to terminate for some arguments in their domain. VDM addresses partial functions by employing nonstandard logical operators, namely the logic of partial functions (LPFs) which can deal with undefined operands. This was developed by Cliff Jones [Jon:90] and is discussed in detail later in the chapter.

Undefined values that arise in Boolean expression are handled by introducing rules to deal with undefined terms: e.g., $T \vee \perp = \perp \vee T = true$; i.e., the truth value of a logical *or* operation is true if at least one of the logical operands is true, and the undefined term is treated as a don't care value. The similarities and differences between Z and VDM (the two most widely used formal methods) are summarized below:

Similarites and Differences of VDM/Z
VDM is a development method including a specification language, whereas *Z* is a specification language only.
Constraints may be placed on types in VDM specifications but not in *Z* specifications.
Z is structured into schemas and VDM into modules.
The schema calculus is part of *Z*.
Relations are part of *Z* but not of VDM.
VDM employs the logic of partial functions (3-valued logic), whereas Z is a classical 2-valued logic.
Preconditions are not separated out in *Z* specifications.

Table 5.1. Similarities and Differences between VDM and *Z*.

EXAMPLE 5.1

The following is a very simple example of a VDM specification and is adapted from [InA:91]. It is a simple library system that allows books to be borrowed and returned. The data types for the library system are first defined and the operation to borrow a book is then defined. It is assumed that the state is made up of three sets and these are the set of books on the shelf, the set of books which are borrowed, and the set of missing books. These sets are mutually disjoint. The effect of the operation to borrow a book is to remove the book from the set of books on the shelf and to add it to the set of borrowed books. The reader is referred to [InA:91] for a detailed explanation.

types
      Bks = Bkd-id-set

state *Library* of
        On-shelf : Bks
        Missing : Bks
        Borrowed : Bks

inv *mk-Library (os, mb, bb)* $\underline{\Delta}$ *is-disj(os,mb,bb)*
end

borrow (b:Bkd-id)
ex wr   *on-shelf, borrowed : Bks*
pre *b* ∈ *on-shelf*
post *on-shelf = on-shelf* $^{\leftarrow}$ *- {b}*   ∧
          borrowed = borrowed $^{\leftarrow}$ ∪ *{b}*

The VDM specification consists of:

- Type definitions
- Stated
- Invariant for the system
- Definition of the operations of the system.

The notation *Bkd-id*-set specifies that *Bks* is a set of *Bkd-ids*; e.g., *Bks* = {$b_1,b_2, ..., b_n$}. The invariant specifies the property that must remain true for the library system: i.e., the sets *on-shelf*, *borrowed*, and *missing* must remain mutually disjoint. The *borrow* operation is defined using preconditions and postconditions. The notation ext wr indicates that the *borrow* operation affects the state, whereas the notation ext rd indicates an operation that does not affect the state. VDM is a widely used formal method and has been used in industrial strength projects as well as by the academic community. These include security-critical systems and safety critical sectors such as the railway industry. There is tool support available, for example, the IFAD VDM-SL[4] toolbox. There are several variants of VDM, including VDM^{++}, an object-oriented extension of VDM, and VDM*, the Irish school of the VDM, which is discussed in the next chapter.

5.2 Sets

Sets are a key building block of VDM specifications. A set is a collection of objects that contains no duplicates. The set of all even natural numbers less than or equal to 10 is given by:

$$S = \{2,4,6,8,10\}.$$

There are a number of in-built sets that are part of VDM including:

Set	Name	Elements
B	Boolean	{true, false}
$\mathbb{N}$	Naturals	{0,1,....}
$\mathbb{N}_1$	Naturals (apart from 0)	{1,2,...}
$\mathbb{Z}$	Integers	{...,-1,0,1,...}
$\mathbb{Q}$	Rational numbers	$\{^p/_q : p,q \in \mathbb{Z} \ q \neq 0\}$
$\mathbb{R}$	Real numbers	

Table 5.2. Built in Types in VDM.

---

[4] As discussed earlier IFAD no longer supplies the VDM tools and CSK in Japan is the new provider of the VDM tools.

The empty set is a set with no members and is denoted by {}. The membership of a set S is denoted by $x \in S$. A set S is a subset of a set T if whenever $x \in S$ then $x \in T$. This is written as $S \subseteq T$. The union of two sets S and T is given by $S \cup T$. The intersection of two sets S and T is given by $S \cap T$.

Sets may be specified by enumeration (as in $S = \{2,4,6,8,10\}$). However, set enumeration is impractical for large sets. The more general form of specification of sets is termed set comprehension, and is of the form:

$$\{\text{set membership} \mid \text{predicate}\}.$$

For example, the specification of the set $T = \{x \in \{2,4,6,8,10\} \mid x > 5\}$ denotes the set $T = \{6,8,10\}$. The set $Q = \{x \in \mathbb{N} \mid x > 5 \wedge x < 8\}$ denotes the set $Q = \{6,7\}$.

The set of all finite subsets of a set $S = \{1,2\}$ is given by:

$$\mathcal{F} S = \{\{\},\{1\},\{2\},\{1,2\}\}.$$

The notation $S : A\text{-set}$ denotes that S is a set, with each element in S drawn from A. E.g., for $A = \{1,2\}$, the valid values of are $S = \{\}$, $S = \{1\}$, $S = \{2\}$, or $S = \{1,2\}$.

The set difference of two sets S and T is given by $S - T$ where:

$$S - T = \{x \in S \mid x \in S \wedge x \notin T\}$$

Given $S = \{2,4,6,8,10\}$ and $T = \{4, 8, 12\}$ then $S - T = \{2,6,10\}$.

Finally, the distributed union and intersection operators are considered. These operators are applied to a set of sets.

$$\cap \{S_1, S_2, \ldots . S_n\} = S_1 \cap S_2 \cap \ldots . \cap S_n$$

$$\cup \{S_1, S_2, \ldots . S_n\} = S_1 \cup S_2 \cup \ldots . \cup S_n$$

The cardinality of a set S is given by card S. This gives the number of elements in the set; for example, card $\{1,3\} = 2$. The notation $Bks = Bkd\text{-}id\text{-set}$ in the Example 5.1 above specifies that Bks is a set of $Bkd\text{-}ids$; e.g., $Bks = \{b_1, b_2, \ldots . b_n\}$.

5.3 Sequences

Sequences are used frequently (e.g., the modeling of stacks via sequences) in VDM specifications. A sequence is a collection of items that are ordered in a particular way. Duplicate items are allowed, whereas in sets they are meaningless. A set may be refined to a sequence of unique elements.

A sequence of elements $x_1, x_2,...,x_n$ is denoted by $[x_1, x_2,...,x_n]$, and the empty sequence is denoted by [ ]. Given a set S, then S^* denotes the set of all finite sequences constructed from the elements of S.

The length of a sequence is given by the *len* operator:

$$len [ ] = 0$$
$$len [1,2,6] = 3.$$

The *hd* operation gives the first element of the sequence. It is applied to non-empty sequences only:

$$hd [x] = x$$
$$hd [x,y,z] = x.$$

The *tl* operation gives the remainder of a sequence after the first element of the sequence has been removed. It is applied to nonempty sequences only:

$$tl [x] = [ ]$$
$$tl [x,y,z] = [y,z]$$

The *elems* operation gives the elements of a sequence. It is applied to both empty and non-empty sequences:

$$elems [ ] = \{ \}$$
$$elems [x,y,z] = \{x, y,z\}.$$

The *indx* operation is applied to both empty and nonempty sequences. It returns the set $\{1,2, ...n\}$ where n is the number of elements in the sequence.

$$inds [ ] = \{ \}$$
$$inds [x,y,z] = \{1, 2, 3\}$$
$$inds \ s = \{1, ... \ len \ s\}.$$

Two sequences may be joined together by the concatenation operator:

$$[ ] \frown [ ] = [ ]$$
$$[x,y,z] \frown [a, b] = [x,y,z, a, b]$$
$$[x,y] \frown [ ] = [x, y].$$

Two sequences s_1 and s_2 are equal if :

$$s_1 = s_2 \Leftrightarrow (len\ s_1 = len\ s_2) \wedge (\forall\ i \in inds\ s_1)\ (s_1\ (i) = s_2(i))$$

Sequences may be employed to specify a stack. For example, a stack of (up to 250) integers is specified as:

state Z-stack of
  stk : $\mathbb{Z}^*$

  inv-Z-stack : $\mathbb{Z}^* \rightarrow \mathbf{B}$
  inv-Z-stack (stk) $\underline{\Delta}$ len stk ≤ 250

  init-mk-Z-stack (stk) $\underline{\Delta}$
        stk = []
end

Table 5.3. Specification of a Stack of Integers.

The push operation is then specified in terms of preconditions/postconditions as follows.

push $(z : \mathbb{Z})$
pre len stk < 100
post stk $= [z] \frown$ stk$^\leftarrow$

5.4 Maps

Maps are employed frequently for modeling in VDM. A map is used to relate the members of two sets X and Y such that each item from the first set X is associated with only one item in the second set Y. Maps are also termed partial functions. The map from X to Y is denoted by:

$$f : T = X \rightarrow^m Y$$

The domain of the map f is a subset of X and the range is a subset of Y. An example of a map declaration is:

$$f : \{Names \rightarrow^m AccountNmr\}.$$

The map f may take on the values:

$$f = \{\ \}$$
$$f = \{eithne \mapsto 231, fred \mapsto 315\}.$$

The domain and range of f are given by:

$$\text{dom } f = \{eithne, fred\}.$$
$$\text{rng } f = \{231, 315\}.$$

The map overwrite operator $f \dagger g$ gives a map that contains all the maplets in the second operand together with the maplets in the first operand that are not in the domain of the second operand.[5]

For $g = \{eithne \mapsto 412, aisling \mapsto 294\}$ then

$$f \dagger g = \{eithne \mapsto 412, aisling \mapsto 294, fred \mapsto 315\}.$$

The map restriction operator[6] has two operands: the first operator is a set, whereas the second operand is a map. It forms the map by extracting those maplets that have the first element equal to a member of the set. For example:

$$\{eithne\} \lhd \{eithne \mapsto 412, aisling \mapsto 294, fred \mapsto 315\} = \{eithne \mapsto 412\}.$$

The map deletion operator has two operands: the first operator is a set, whereas the second operand is a map. It forms the map by deleting those maplets that have the first element equal to a member of the set. For example:

$$\{eithne, fred\} \lhdminus \{eithne \mapsto 412, aisling \mapsto 294, fred \mapsto 315\} = \{aisling \mapsto 294\}$$

Total maps are termed functions, and a total function f from a set X to a set Y is denoted by:

$$f : X \rightarrow Y$$

A partial function (map) is denoted by $f : X \rightarrow^m Y$, and may be undefined for some values in X. Partial maps arise frequently in specifications, as often in practice a function will be undefined for one or more values in its domain. For example, the function $f(x) = 1/x$ is undefined for $x = 0$. Consequently, if $1/0$ arises in an expression, then that expression is undefined.

----

[5] $f \dagger g$ is the VDM notation for function override. The notation $f \oplus g$ is employed in Z.

[6] The map restrictor and map deletion operators are similar to the Z domain restrictor and anti-restriction operators.

5.5 Logic in VDM

Logic has been discussed in detail earlier in this book. This section discusses the logic of partial functions (LPFs) used in VDM to deal with terms that may be undefined. It was developed by Cliff Jones [Jon:90], and is a 3-valued logic where a logical term may be true, false, or undefined. The truth functional operators in this 3-valued logic are:

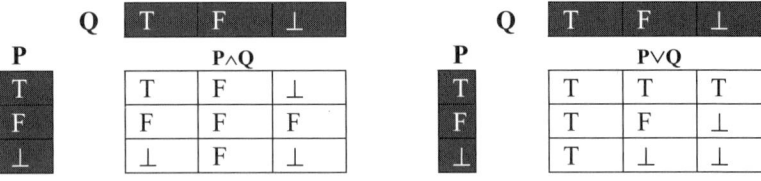

Q	T	F	⊥		Q	T	F	⊥
P		P∧Q			**P**		P∨Q	
T	T	F	⊥		T	T	T	T
F	F	F	F		F	T	F	⊥
⊥	⊥	F	⊥		⊥	T	⊥	⊥

Fig. 5.1. Conjunction   **Fig. 5.2.** Disjunction

The conjunction of P and Q is true when both P and Q are true; false if one of P or Q is false, and undefined otherwise. The operation is commutative. The disjunction of P and Q (P ∨ Q) is true if one of P or Q is true; false if both P and Q are false; and undefined otherwise. The implication operation (P ⇒ Q) is true when P is false or when Q is true; it is undefined otherwise[7].

Q	T	F	⊥		Q	T	F	⊥
P		P⇒Q			**P**		P≡Q	
T	T	F	⊥		T	T	F	⊥
F	T	T	T		F	F	T	⊥
⊥	T	⊥	⊥		⊥	⊥	⊥	⊥

Fig. 5.3. Implication   **Fig. 5.4.** Equivalence

The equivalence operation (P≡Q) is true when both P and Q are true or false; it is false when P is true and Q is false (or vice versa); and it is undefined otherwise. The not operator (¬) is a unary operator such ¬A is true when A is false, false when A is true, and undefined when A is undefined.

A	¬A
T	F
F	T
⊥	⊥

Fig. 5.5. Disjunction

---

[7] The problem with 3-valued logic is that they are less intuitive than classical 2-valued logic.

It is clear from the truth table definitions that the result of the operation may be known immediately after knowing the value of one of the operands (e.g., disjunction is true if P is true irrespective of the value of Q). The law of the excluded middle: i.e., $A \vee \neg A$ = true does not hold in the 3-valued logic of partial functions. However, this is reasonable [Jon:90] as one would not expect the following to be true:

$$(\tfrac{1}{0} = 1) \vee (\tfrac{1}{0} \neq 1).$$

5.6 Data Types and Data Invariants

Larger specifications require more complex data types. The VDM specification language allows composite data types to be created from their underlying component data types. For example, the composite data type *Date* is defined as follows [InA:91]:

Date = compose *Date* of
         year : {2000,...,3000}
         month {1,...,12}
         day : {1,...,31}
         end

Table 5.4. Composite Data Types in VDM.

A make function is employed to construct a date from the components of the date; i.e., the *mk-Date* function takes three numbers as arguments and constructs a date from them.

$$mk\text{-}Date : \{2000,...,3000\} \times \{1,...,12\} \times \{1,...,31\} \rightarrow Date$$

For example, the date of 5th August 2004 is constructed as follows:

$$mk\text{-}Date \ (2004, 8, 5).$$

Selectors are employed to take a complex data type apart into its components. The selectors employed for date are day, month, year. Hence, the selection of the year component in the date of 5th August 2004 is:

$$mk\text{-}Date : (2004, 8, 5). \ year = 2004.$$

The observant reader will note that the definition of the *Date* data type above allows invalid dates to be present: e.g., 29th February 2001 and 31st November 2004. Hence, what is required is a predicate to restrict elements of the data type to be valid dates. This is achieved by a data invariant.

$$
\begin{array}{l}
\textit{Inv-Date} : \textit{Date} \rightarrow \mathbf{B} \\
\textit{Inv-Date} \ (dt) \ \underline{\Delta} \\
\qquad \text{let } \textit{mk-Date} \ (yr, md, dy) = dt \text{ in} \\
\qquad\qquad (md \in \{1,3,5,7,8,10,12\} \wedge dy \in \{1,...,31\}) \\
\qquad\quad \vee \ (md \in \{4,6,9,11\} \wedge dy \in \{1,...,30\}) \\
\qquad\quad \vee \ (md = 2 \wedge \textit{isleapyear}(yr) \wedge dy \in \{1,...,29\}) \\
\qquad\quad \vee \ (md = 2 \wedge \neg\textit{isleapyear}(yr) \wedge dy \in \{1,..,28\})
\end{array}
$$

Table 5.5. Composite Data Invariant for Composite Date Data Type.

Any operation that affects the date will need to preserve the data invariant. This gives rise to a proof obligation for each operation that affects the date.

5.7 Specification in VDM

An abstract machine (sometimes called object) consists of the specification of a data type together with the operations on the data type. The production of a large specification involves [InA:91]:

1. Identifying and specifying the abstract machines.
2. Defining how these machines fit together and are controlled to provide the required functionality.

The abstract machines may be identified using design tools such as data flow diagrams and object-oriented design.[8] Once the abstract machines have been identified there are then two further problems to be addressed.

1. How are the abstract machines to be used?
   (e.g., users or other programs?).
2. How are the abstract machines to be implemented in code?[9]

VDM-SL specifications are like programs except they are not executable. However, one important difference is that there are no side effects in VDM-SL expressions. Side effects are common in imperative programming languages such as C or Pascal due to the use of the assignment statement which has the effect of modifying the values of the variables. Functional programming languages such as Miranda or Haskell do not employ assignment and therefore there are no side effects. This means that the value of an expression remains the same throughout the execution of the program, whereas this is not the case in C

---

[8] Other approaches would be to use some of the ideas of Parnas on module decomposition or to employ methods such as UML to identify the classes required.

[9] The implementation in code generally requires refinement of the abstract machine into more concrete machines until the executable code is eventually produced. This is discussed in the next section.

or Pascal. The VDM-SL specification language does not contain side effects and so the value of an expression remains constant throughout the specification.

The VDM specification is structured into type definitions, state definitions, an invariant for the system, the initial state, and the definition of the operations of the system. These are described in the table below:

Name	Description
Type definitions	The type definitions specify the data types employed. These include the built-in sets, or sets constructed from existing sets. A domain-invariant predicate may further constrain the definition. A type in VDM is more specific than the signature of the type, and represents values in the domain defined by the signature, which satisfy the domain invariant.
State definitions	This is the definition of the collection of stored data. The operations access/modify the data.
(Data-)Invariant for the system	This describes a condition that must be true for the state throughout the execution of the system.
Initial value of the state	This specifies the initial value of the state.
Definition of operations	The operations on the state are defined in terms of preconditions and postconditions. The keywords **rd** and **wr** indicate whether the operation changes the state.

Table 5.6. Structure of VDM Specification.

The whole of the development process is based on the formal specification, and it is therefore essential that the specification is correct. It is important in a sense to prove that various properties are true of the specification and to thereby provide confidence that the specification is correct. A description of the development of the specification of the library system is presented in [InA:91].

5.8 Refinement

The development of executable code from a VDM specification involves breaking down the specification into smaller specifications (each smaller specification defines an easier problem) [InA:91]. Each smaller specification is then tackled (this may involve even smaller subspecifications) until eventually the implementation of the code that satisfies each smaller specification is trivial as are the corresponding proofs of correctness. The code fragments are then glued together using the programming language constructs of the semicolon, the conditional statement, and the while loop.

At each step of the process a proof of correctness is conducted to ensure that the refinement is valid. The approach allows a large specification to be broken down to a smaller set of specifications that can be translated into code. The approach involves deriving equivalent specifications to existing specifications. A specification OP' is equivalent to a given specification OP if any program that satisfies OP' also satisfies OP. The formal definition of equivalence is:

1. $\forall i \in$ State . pre-$Op(i) \Rightarrow$ pre-$OP'(i)$
2. $\forall i,o \in$ State . pre-$Op(i) \wedge$ post-$Op'(i,o) \Rightarrow$ post-$OP(i,o)$

The idea of a program satisfying its specification can be expanded to a specification satisfying a specification as follows:

OP' sat OP if
1. $\forall i \in$ State . pre-$Op(i) \Rightarrow$ pre-$OP'(i)$
2. $\forall i,o \in$ State . pre-$Op(i) \wedge$ post-$Op'(i,o) \Rightarrow$ post-$OP(i,o)$
3. $\forall i \in$ State . pre-$Op'(i) \Rightarrow \exists o \in$ State . post-$OP'(i,o)$.

The formal definition requires that whenever an input satisfies the precondition of *OP*, then it must also satisfy the precondition of *OP'*. Further, the two specifications must agree on an answer for any input state variables that satisfy the precondition for *OP*. Finally, the third part expresses the idea of a specification terminating (similar to a program terminating). It expresses the requirement that the specification is implementable.

The production of a working program that satisfies the specification is evidence that a specification is satisfiable. There is a danger that the miracle program could be introduced while carrying out a program refinement. The miracle program is a program that has no implementable specification:

> **miracle**
> pre   true
> post   false

Clearly, an executable version of miracle is not possible as the miracle program must be willing to accept any input and produce no output. Refinement is a weaker form of satisfaction (and allows the *miracle* program). It is denoted by the $\sqsubseteq$ operator.

$$A \text{ sat } B \Rightarrow B \sqsubseteq A$$
$$A \sqsubseteq B \text{ and } B \text{ is implementable} \Rightarrow B \text{ sat } A$$

$$S \sqsubseteq R_1 \sqsubseteq R_2 \sqsubseteq \ ... \sqsubseteq R_n \sqsubseteq p \ \wedge \ p \text{ is executable}$$
$$\Rightarrow \ p \text{ sat } S$$

5.9 Tools for VDM

Various tools (academic and commercial) have been developed for VDM. These tools include syntax checkers to check the syntactic validity of the formal specification, specialized editors, tools to support refinement, and code generators to generate a high-level programming language from the formal specification, and theorem provers that are employed to assist with proof.

The earliest tools for VDM were academic rather than commercial and this includes work done at Manchester University by Cliff Jones and others on the Mural tool. The development of the Mural system[10] involved the collaboration of Rutherford Laboratories and ICL as industrial partners.

The main part of the Mural system is a proof assistant that supports reasoning about formal specifications. It allows the various proof obligations associated with a formal specification to be carried out thereby ensuring the internal consistency of the specification. Mural also contains a VDM support tool to construct specifications and support refinement. This is done using the built-in structure editor for writing VDM specifications or reading a file generated by the Adelard Specbox tool (that has been syntactically checked). Mural has a good user interface which aims to make the interaction with the system as easy as possible.

The SpecBox tool has been developed by Adelard. This tool provides a syntax checking tool of VDM specifications and also provides simple semantic checks. It allows the facility to generate a LaTeX file to enable the specifications to be printed in mathematical notation. Finally, it includes a translator to the Mural proof assistant.[11]

Perhaps, the most widely known VDM support tool is the IFAD VDM-SL Toolbox (now renamed to VDMTools). This tool was originally developed by IFAD based in Odense, Denmark, but IFAD has recently sold the Toolbox to

---

[10] The kernel of Mural was specified in VDM and is an example of the formal methods community taking its own medicine. However, the Mural tool was never commercialized and it appears that the work in Manchester University and Appleton is no longer taking place. I have heard (from John Fitzgerald) that the Centre for Software Reliability (CSR) plans to put the Mural book on-line.

[11] The tool needs to be updated to be compatible with the ISO/IEC 13817-1 standard for VDM, as it is based on an earlier draft of the standard. I am unclear as to how widely the SpecBox tool is used.

the CSK Group of Japan (an information technology company employing about 10,000 people worldwide). CSK plans to make the tools more widely available and further details are available from VDM_SP@cii.csk.co.jp or via the VDM web inquiry form at (https://www.csk.co.jp/support_e/vdm.html) on the CSK web site.

The original IFAD Toolbox provided syntax and semantic checking of VDM specifications, LaTeX pretty printing, and a code generator to convert from VDM-SL to C++. It was then extended to provide support to both VDM-SL and VDM++ (the object-oriented extension to VDM). The main features of the VDMTools for VDM++ are:

Functionality	Description
Specification manager	Keeps track of status of classes in the specification.
Syntax checker	Checks whether VDM++ syntax is correct.
Type checker	Identifies misuses of values and operators.
Interpreter and debugger	Allows execution of executable constructs in VDM++. This provides a running prototype. The Debugger allows break-points to be set and inspection of variables within scope.
Integrity Examiner	This includes checks for integrity violations (e.g., violations of invariants, preconditions, and postconditions).
Test facility	This allows the execution of a suite of test cases and test coverage information may be recorded.
Automatic code generator	Automatic generation of C++ and Java code from VDM++ specifications (for 95% of all VDM++ constructs). The remaining 5% requires user-defined code for the nonexecutable parts of the specification.
Corba compliant API	This allows other programs to access a running toolbox through a Corba compliant API.
Rose to VDM++ link	This provides a bidirectional link between the Toolbox and Rational Rose thereby providing a bidirectional link between VDM++ and UML.[12]
Java to VDM++ trans-	This allows existing Java applications

---

[12] The Rational Rose tool is becoming a legacy tool as the Rational Software Architect (RSA) tool and Requisite Pro tool are the next generation of UML tools from IBM/Rational.

lator	to be reverse engineered to VDM++. This is useful if new development needs to take place on legacy Java software.

Table 5.7. Functionality of VDM Tools.

Further information on the VDM tools is available on (http://www.vdmbook.com/tools.php). The Centre for Software Reliability in the United Kingdom (http://www.csr.ncl.ac.uk/) may also be contacted for up to date information on the VDM tools.

Finally, there is an open source project aimed at generating new generation tools for VDM++ that has recently commenced. Further information on this open source project is available at http://www.overturetool.org/.

5.10 Summary

VDM dates from work done by the IBM research laboratory in Vienna in the 1960s. It includes a formal specification language (originally called Meta IV) but recently standardized as an ISO standard (International Standard ISO/IEC 13817-1). VDM includes a formal specification language and a method to develop high-quality software. The Vienna group was broken up in the mid-1970s and this led to the formation of different schools of the VDM in diverse locations. Further work on VDM and Meta-IV continued in the 1980s and standards for VDM (VDM-SL) appeared in the 1990s.

It is a *model-oriented approach*, which means that an explicit model of the state of an abstract machine is given, and operations are defined in terms of this state. Operations are defined in a precondition and postcondition style. Each operation has an associated proof obligation to ensure that if the precondition is true, then the operation preserves the system invariant. VDM employs postconditions to stipulate the effect of the operation on the state. The postcondition specifies the new state using a predicate that relates the prestate to the poststate.

VDM is both a specification language and a development method. Its method provides rules to verify the steps of development and enable the executable specification, i.e., the detailed code, to be obtained from the initial specification via refinement steps:

$$S = S_0 \sqsubseteq S_1 \sqsubseteq S_2 \sqsubseteq \ ... \sqsubseteq\ S_n = E$$

Retrieval functions enable a return from a more concrete specification, to the more abstract specification. The initial specification consists of an initial state, a system state, and a set of operations.

VDM specifications are structured into modules, with a module containing the module name, parameters, types, and operations. VDM employs the logic of partial functions (LPFs) to deal with undefined operands. VDM has been used in industrial-strength projects as well as by the academic community.

There is tool support available, for example, the IFAD VDM-SL toolbox. There are several variants of VDM, including VDM^{++}, an object-oriented extension of VDM, and VDM$^{\clubsuit}$, the Irish school of the VDM, which is discussed in the next chapter.

6
Irish School of VDM

6.1 Introduction

The Irish School of VDM is a variant of standard VDM, and is characterized by [Mac:90][1] its constructive approach, classical mathematical style, and its terse notation. In particular, this method combines the *what* and *how* of formal methods in that its terse specification style stipulates in concise form *what* the system should do; furthermore, the fact that its specifications are constructive (or functional) means that that the *how* is included with the *what*. However, it is important to qualify this by stating that the how as presented by VDM* is not directly executable, as several of its mathematical data types have no corresponding structure in high-level programming languages or functional languages. Thus a conversion or reification of the specification into a functional or higher-level language must take place to ensure a successful execution. It should be noted that the fact that a specification is constructive is no guarantee that it is a good implementation strategy, if the construction itself is naive. This issue is considered (cf. pp. 135-7 in [Mac:90]), and the example considered is the construction of the Fibonacci series.

   The Irish school follows a similar development methodology as in standard VDM and is a model-oriented approach. The initial specification is presented, with initial state and operations defined. The operations are presented with preconditions; however, no postcondition is necessary as the operation is "functionally" i.e., explicitly constructed. Each operation has an associated proof obligation; if the precondition for the operation is true and the operation is performed, then the system invariant remains true after the operation. The proof of invariant preservation normally takes the form of *constructive proofs*. This is especially the case for *existence proofs*, in that the philosophy of the school is to

---

[1] This chapter is dedicated to Dr. Mícheal Mac An Airchinnigh of Trinity College, Dublin who founded the Irish school of VDM. Mícheal is a former Christian brother and has great wit and charisma. He is also an excellent orator and I remember a very humorous after dinner speech made in Odense, Denmark at the FME'93 conference. In the speech, he explained why he was Aristotelian and said: "Why am I Aristotelian? Well, my wife is and I can't be Platonic with her." He also staged an alternate version of the Emperors New Clothes as part of the entertainment at FME'93. This included participation from Michael Butler, Andrew Butterfield, and Eoin McDonnell.

go further than to provide a theoretical proof of existence; rather the aim is to exhibit existence constructively.

The emphasis is on constructive existence and the implication of this is that the school avoids the existential quantifier of predicate calculus. In fact, reliance on logic in proof is kept to a minimum, and emphasis instead is placed on equational reasoning rather than on applying the rules of predicate calculus. Special emphasis is placed on studying algebraic structures and their morphisms. Structures with nice algebraic properties are sought, and such a structure includes the monoid, which has closure, associativity, and a unit element. The monoid is a very common structure in computer science, and thus it is appropriate to study and understand it. The concept of isomorphism is powerful, reflecting that two structures are essentially identical, and thus we may choose to work with either, depending on which is more convenient for the task at hand.

The school has been influenced by the work of Polya and Lakatos. The former [Pol:57] advocated a style of problem solving characterized by solving a complex problem by first considering an easier subproblem and considering several examples, which generally leads to a clearer insight into solving the main problem. Lakatos's approach to mathematical discovery (cf. [Lak:76]) is characterized by heuristic methods. A primitive conjecture is proposed and if global counterexamples to the statement of the conjecture are discovered, then the corresponding "hidden lemma" for which this global counterexample is a local counterexample is identified and added to the statement of the primitive conjecture. The process repeats, until no more global counterexamples are found. A skeptical view of absolute truth or certainty is inherent in this.

Partial functions are the norm in VDM$^\clubsuit$, and as in standard VDM, the problem is that recursively defined functions may be undefined, or fail to terminate for several of the arguments in their domain. The logic of partial functions (LPFs) is avoided, and instead care is taken with recursive definitions to ensure termination is achieved for each argument. This is achieved by ensuring that the recursive argument is strictly decreasing in each recursive invocation. The $\perp$ symbol is typically used in the Irish school to represent *undefined or unavailable* or *do not care*. Academic and industrial projects[2] have been conducted using the method of the Irish school, but at this stage tool support is limited.

EXAMPLE

The following is the equivalent VDM$^\clubsuit$ specification of the earlier example of a simple library presented in standard VDM.

$Bks = \mathbb{P} \ Bkd\text{-}id$
$Library \ = (Bks \times Bks \times Bks)$

---

[2] This includes practical industrial work with Motorola in the US and participation in the EU SCORE project (part of the RACE II programme). The latter included work in feature interaction modeling and detection which is important in the telecoms sector.

Os, Ms, Bw ∈ *Bks*

inv- *Library (Os, Ms, Bw)* $\underline{\Delta}$ *Os* ∩ *Ms* = Ø

$\qquad\qquad\qquad\qquad$ ∧ *Os* ∩ *Bw* = Ø

$\qquad\qquad\qquad\qquad$ ∧ *Bw* ∩ *Ms* = Ø

Bor: Bkd-id → (*Bks* × *Bks*) → (*Bks* × *Bks*)

Bor ⟦*b*⟧ (*Os, Bw*) $\underline{\Delta}$ (⊲ ⟦*b*⟧ *Os*, *Bw* ∪{*b*})

pre-*Bor: Bkd-id* → (*Bks* × *Bks*) → **B**

pre-*Bor* ⟦*b*⟧ *(Os, Bw)* $\underline{\Delta}$ χ ⟦*b*⟧ *Os*

There is, of course, a proof obligation to prove that the *Bor* operation preserves the invariant, i.e., that the three sets of borrowed, missing, or on the shelf remain disjoint after the execution of the operation. Proof obligations require a mathematical proof by hand or a machine-assisted proof to verify that the invariant remains satisfied after the operation.

$$\text{pre-}Bor \; ⟦b⟧ \; (Os, Bw) \wedge ((Os', Bs') = Bor \; ⟦b⟧ \; (Os, Bw))$$
$$\Rightarrow \text{inv-}Library \; (Os', Ms', Bw')$$

6.2 Mathematical Structures and Their Morphisms

The Irish school of VDM uses mathematical structures to organize the modeling of systems and to organize proofs. There is an emphasis on identifying useful structures that will assist modeling and constructing new structures from existing ones. Some well-known structures used in VDM* include semigroups and monoids

A semigroup is a structure A with a binary operation * such that the closure and associativity properties hold:

$$a * b \in A \qquad\qquad \forall \, a, b \in A$$
$$(a * b) * c = a * (b * c) \;\; \forall \, a, b, c \in A$$

Examples of semigroups include the natural numbers under addition, nonempty sets under the set union operation, and nonempty sequences under concatenation. A semi-group is commutative if:

$$a * b = b * a \qquad\qquad \forall \, a, b \in A.$$

A monoid *M* is a semigroup that has the additional property that there is an identity element *u* ∈ *M* such that:

$$a * b \in M \qquad\qquad \forall \, a, b \in M$$

$$(a * b) * c = a * (b * c) \quad \forall \, a, \, b \,, c \in M$$
$$a * u = a = u * a \qquad \forall \, a \in M$$

Examples of monoids include the integers under addition, sets under the set union operation, and sequences under concatenation. The identity element is 0 for the integers, the empty set $\varnothing$ for set union, and the empty sequence Λ for sequence concatenation. A monoid is commutative if $a * b = b * a \; \forall \, a,b \in M$. A monoid is denoted by $(M, *, u)$.

A function $h : (M, \oplus, u) \rightarrow (N, \otimes, v)$ is structure preserving (morphism) between two monoids $(M, \oplus, u)$ and $(N, \otimes, v)$ if the same result is obtained by either:

1. Evaluating the expression in M and then applying h to the result.
2. Applying h to each element of M and evaluating the result under $\otimes$.

A monoid homomorphism $h : (M, \oplus, u) \rightarrow (N, \otimes, v)$ is expressed in the commuting diagram below. It requires that the commuting diagram property holds and that the image of the identity of M is the identity of N.

$$h(m_1 \oplus m_2) = h(m_1) \otimes h(m_2) \qquad \forall m_1, m_2 \in M$$
$$h(u) = v$$

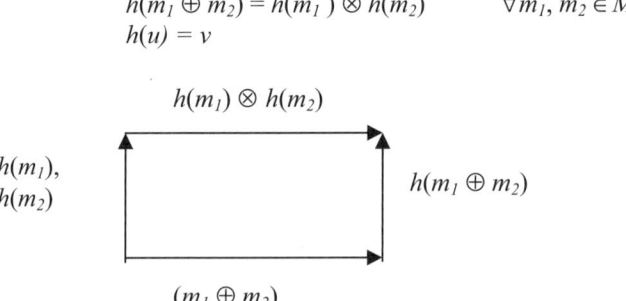

$$h(m_1) \otimes h(m_2)$$

$$h(m_1), \\ h(m_2)$$

$$h(m_1 \oplus m_2)$$

$$(m_1 \oplus m_2)$$

Fig. 6.1. Monoid Homomorphism

A morphism $h : (M, \oplus, u) \rightarrow (M, \oplus, u)$ is termed an endomorphism.

EXAMPLES

Consider the monoid of sequences $(\Sigma^*, \cap, \Lambda)^3$ and the monoid of natural numbers $(\mathbb{N}, +, 0)$. Then the function *len* that gives the length of a sequence is a monoid homomorphism from $(\Sigma^*, \cap, \Lambda)$ to $(\mathbb{N}, +, 0)$. Clearly, $len(\Lambda) = 0$ and the commuting diagram property holds:

---

[3] The Irish VDM notation includes sets, sequences and functions and is described later in this chapter. One striking feature is its use of the Greek alphabet, and the above defines the monoid of sequences over the alphabet Σ. The concatenation operator is denoted by $\cap$ and the empty sequence is denoted by Λ. The Greek influence is also there in the emphasis on constructive proofs such as the proofs in Euclidean geometry whereby the existence of an object is demonstrated by its construction.

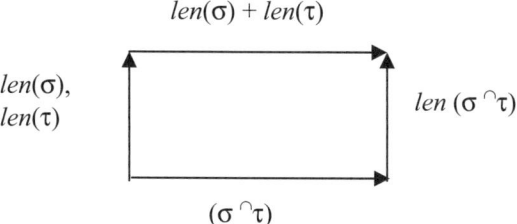

Fig. 6.2. *Len* Homomorphism.

The second example considered is from the monoid of sequences to the monoid of sets under set union. Then the function *elems* that gives the elements of a sequence is a monoid homomorphism from $(\Sigma^*, \frown, \Lambda)$ to $(\mathbb{P}\Sigma, \cup, \varnothing)$. Clearly, $elems(\Lambda) = \varnothing$ and the commuting diagram property holds.

Consider the set removal operation $\lessdot \llbracket S \rrbracket$ on the monoid of sets under set union. Then the removal operation is a monoid endomorphism from $(\mathbb{P}\Sigma, \cup, \varnothing)$ to $(\mathbb{P}\Sigma, \cup, \varnothing)$.

$$\lessdot \llbracket S \rrbracket \, (S_1 \cup S_2) = \lessdot \llbracket S \rrbracket \, (S_1) \cup \lessdot \llbracket S \rrbracket \, (S_2) \qquad \forall S_1, S_2 \subseteq S$$
$$\lessdot \llbracket S \rrbracket \, (\varnothing) = \varnothing$$

$$\lessdot \llbracket S \rrbracket \, (S_1) \cup \lessdot \llbracket S \rrbracket \, (S_2)$$

$\lessdot \llbracket S \rrbracket \, (S_1)$
$\lessdot \llbracket S \rrbracket \, (S_2)$ $\lessdot \llbracket S \rrbracket \, (S_1 \cup S_2)$

$$(S_1 \cup S_2)$$

Fig. 6.3. Set Removal Endomorphism.

Set restriction $(\lhd \, \llbracket S \rrbracket)$ is also an endomorphism on $(\mathbb{P}\Sigma, \cup, \varnothing)$.

COMMENT (MONOIDS AND MORPHISMS)
Monoids and their morphisms are useful and are used widely in VDM$^\clubsuit$. They are well-behaved structures and allow compact definitions of functions and also simplify proof. The use of monoids and morphisms helps to produce compact models and proofs.

6.3 Models and Modeling

A model is a representation of the physical world. However, the model is not the reality, but rather a simplification of the reality. Therefore, models do not include all aspects of the reality, but it is important that the model includes all essential aspects of the reality. Models are generally mathematical representations of the physical world.

The adequacy of a model is a key concern. For example, the model of the Tacoma Narrows Bridge[4] did not include aerodynamic forces, and this inadequacy in the model had a major influence on the eventual collapse of the bridge. It is essential to explore the consequences of a model, and to determine if the model is an adequate representation of the reality. Occasionally, there may be more than one model to explain the reality. For example, Ptolemy's Cosmological Model and the Copernican Model. Both models are adequate at explaining aspects of navigation. In fact, the Copernican model was less accurate than Ptolemy's model, until the former was revised by Kepler [Kuh:70]. Occam's Razor (or the *Principle of Parsimony*) is a key principle underlying modeling. The principle is stated as follows: *Entia non sunt multiplicanda praeter necessitatem*; this essentially means that the number of entities required to explain anything should be kept to a minimum. The implication of this principle is that the modeler should seek the simplest model with the least number of assumptions. The principle is attributed to the medieval philosopher William of Occam[5].

The key application of Occam's Razor in practice is to remove all superfluous concepts which are not needed to explain the phenomenon. The net result is a crisp and simpler model. In theory, this should reduce the likelihood of introducing inconsistencies and errors into the model. Such a model captures the essence of the reality.

In summary, a model is an abstraction or simplification of reality. Model exploration enables an informed decision to be made on the adequacy or otherwise of the model. The model should be kept as simple as possible. The model itself then serves as a formal means of testing hypotheses or ideas about some aspects of the world. This involves the formulation of questions which are then answered in terms of the model. Mathematical models offer a valuable means for the examination of aspects of the world.

6.4 Sets

Sets have been discussed earlier in the book and this section focuses on their use in VDM*. A set is a collection of objects all drawn from the same domain. Sets

---

[4] The Tacoma Narrows Bridge (known as Galloping Gertie) collapsed in 1940 due to a design flaw. Further details are in [ORg:02].

[5] William of Occam was a medieval philosopher and theologian. Occam is a village in Surrey, England.

may be enumerated by listing all of their elements. Thus, the set of all even natural numbers less than or equal to 10 is:

$$\{2,4,6,8,10\}.$$

The membership of a set S is denoted by $x \in S$. There is also another notation for set membership based on the characteristic function:

$$\chi : \Sigma \rightarrow \mathbb{P}\Sigma \rightarrow \mathbf{B}$$
$$\chi[\![x]\!] \, S \, \underline{\Delta} \, x \in S.$$

The empty set is denoted by $\emptyset$. Various operations on sets such as union, intersection, difference, and symmetric difference are employed. The union of two sets S and T is given by $S \cup T$, and their intersection by $S \cap T$. The set restriction operation for S on T restricts T to its elements that are in S; it is given by:

$$\triangleleft \; [\![S]\!] \; T = S \cap T.$$

This is also written in infix form as:

$$S \triangleleft T = S \cap T.$$

The set difference (or set removal operation) of two sets S, T is given by $S \setminus T$. It is also written as:

$$\triangleleft \; [\![T]\!] \; S = S \setminus T$$

or in infix form as:

$$T \triangleleft S = S \setminus T.$$

The symmetric difference operation is given by:

$$S \Delta T \; \underline{\Delta} \; (S \cup T) \setminus (S \cap T).$$

The number of elements in a set S is given by the cardinality function $card(S)$.

$$card(S) = \#S = |S|.$$

The powerset of a set X is the set of all subsets of X. It is denoted by $\mathbb{P}X$ and includes the empty set. The notation $\mathbb{P}'Y$ denotes the set of non-empty subsets of Y: i.e., $\triangleleft [\![\emptyset]\!] \; \mathbb{P}Y$.

The set S is said to be a subset of T ($S \subseteq T$) if whenever $s \in S$ then $s \in T$. The distributed union of set of sets is defined as:

$$\cup/ \{S_1, S_2, \ldots, S_n\} = S_1 \cup S_2 \cup \ldots \cup S_n$$

6.5 Relations and Functions

There is no specific notation for relations in VDM$^{\clubsuit}$. Instead, relations from a set X to a set Y are modeled by either:

- $R \subseteq \mathbb{P}\ (X \times Y)$
- A partial functions ρ of the form ρ: $X \rightarrow \mathbb{P}\ ' Y$.

An element *x* is related to *y* if:

- $(x,y) \in R$

  or

- $\chi[\![x]\!]\ \rho \wedge y \in \rho(x)$.

The structure $(X \rightarrow \mathbb{P}\ ' Y)$ is isomorphic to $\mathbb{P}\ (X \times Y)$.

The functional view of relations uses the indexed monoid $(X \rightarrow \mathbb{P}\ 'Y, \odot, \theta)$ and this allows the familiar relational operations such as relational inverse, relational composition, etc., to be expressed functionally. For example, the inverse of a relation ρ: $(X \rightarrow \mathbb{P}\ ' Y)$ is of the form ρ^{-1}: $(Y \rightarrow \mathbb{P}\ 'X)$, and the relation inverse may be defined constructively.

A function takes values from one domain and returns results in another domain. The map $\mu : X \rightarrow Y$ denotes a partial function from the set *X* to the set *Y*. The result of the function for a particular value *x* in the domain of μ is given by μ (x). The empty map from *X* to *Y* is denoted by θ.

The domain of a map μ is given by dom μ, and it gives the elements of *X* for which the map μ is defined. The notation $x \in$ dom μ indicates that the element *x* is in the domain of μ. This is often written with a slight abuse of notation as $x \in \mu$. Clearly, dom θ = ∅ and dom $\{x \rightarrow y\} = \{x\}$. The domain of μ is *X* if μ is total.

New maps may be constructed from existing maps using function override. The function override operator was defined in *Z* and the operator combines two functions of the same type to give a new function of the same type. The effect of the override operation $(\mu \dagger v)$ is that an entry $\{x \mapsto y\}$ is removed from the map μ and replaced with the entry $\{x \mapsto z\}$ in *v*.

The notion $f \oplus g$ is employed for function override in *Z*. It indicates that *f* is overridden by *g*. The notation $(\mu \dagger v)$ is employed for function override in VDM$^{\clubsuit}$.

$$(\mu \dagger v)\ (x) = v\ (x) \text{ where } x \in \text{dom } v$$

$$(\mu \dagger v)\ (x) = \mu\ (x) \text{ where } x \notin \text{dom } v \wedge x \in \text{dom } \mu$$

Maps under override form a monoid $(X \to Y, \dagger, \theta)$ with the empty map θ the identity of the monoid. The domain (dom) operator is a monoid homomorphisms. The domain homomorphism is of the form:

$$\text{dom} : (X \to Y, \dagger, \theta) \to (\mathbb{P}X, \cup, \varnothing).$$

$$\text{dom}\ \{x \mapsto y\} = \{x\}$$

Domain removal and domain restriction operators were discussed for sets in the previous section. The domain removal operator ($\triangleleft\!\![S]$) and the domain restriction operator ($\triangleleft [|S|]$) are endomorphisms of $(X \to Y, \dagger, \theta)$. The domain removal operator ($\triangleleft\!\![S]$) is defined as follows:

$$\triangleleft\!\![S] : (X \to Y, \dagger, \theta) \to (X \to Y, \dagger, \theta)$$

$$\triangleleft\!\!|S|\ \{x \mapsto y\} \underset{\Delta}{} \theta \qquad\qquad (x \in S)$$
$$\triangleleft\!\![S]\ \{x \mapsto y\} \underset{\Delta}{} \{x \mapsto y\} \qquad (x \notin S)$$

The domain restriction operator ($\triangleleft[|S|]$) is defined as follows:

$$\triangleleft [|S|] : (X \to Y, \dagger, \theta) \to (X \to Y, \dagger, \theta)$$

$$\triangleleft [|S|]\ \{x \mapsto y\} \underset{\Delta}{} \{x \mapsto y\}\ (x \in S)$$
$$\triangleleft [|S|]\ \{x \mapsto y\} \underset{\Delta}{} \theta \qquad\quad (x \notin S)$$

The restrict and removal operators are extended to restriction/removal from another map by abuse of notation:

$$\triangleleft [|\mu|]\ v = \triangleleft [|\text{dom}\ \mu|]\ v$$
$$\triangleleft\!\![|\mu|]\ v = \triangleleft\!\![|\text{dom}\ \mu|]\ v.$$

Given an **injective** total function $f : (X \to W)$ and a total function $g : (Y \to Z)$ then the map functor $(f \to g)$ is a homomorphism of

$$(f \to g) : (X \to Y, \dagger, \theta) \to (W \to Z, \dagger, \theta)$$
$$(f \to g)\ \{x \mapsto y\} = \{f(x) \mapsto g(y)\}.$$

Currying is used extensively in VDM⁺ and the term is named after the logician Haskell Curry.[6] Currying involves replacing a function of n arguments by the application of n functions of 1-argument. Consider the function $f : X \times Y \rightarrow Z$. Then the usual function application is:

$$f(x,y) = z.$$

The curried form of the above is application is:

$$f : X \rightarrow Y \rightarrow Z$$

$$f \llbracket x \rrbracket \text{ is a function} : Y \rightarrow Z \text{ and } f \llbracket x \rrbracket \, y = z$$

6.6 Sequences

Sequences are ordered lists of zero or more elements from the same set. The set of sequences from the set Σ is denoted by Σ^*, and the set of nonempty sequences is denoted by Σ^+. Two sequences σ and τ are combined by sequence concatenation to give $\sigma \frown \tau$. The structure $(\Sigma^*, \frown, \Lambda)$ is a monoid under sequence concatenation, and the identity of the monoid is the empty sequence Λ.

The sequence constructor operator (-:-) takes an element x from the set Σ, and a sequence σ from Σ^*, and produces a new sequence σ' that consists of the element x as the first element of σ' and with the remainder of the sequence given by σ.

$$\sigma' = x : \sigma$$

The most basic sequence is given by:

$$\sigma = x : \Lambda$$

A sequence constructed of n elements $x_1, x_2, \ldots x_{n-1}, x_n$ (in that order) is given by:

$$x_1 : (x_2 : \ldots : (x_{n-1} : (x_n : \Lambda))\ldots).$$

This is also written as:

$$\langle x_1, x_2, \ldots x_{n-1}, x_n \rangle$$

The head of a nonempty sequence is given by:

$$\text{hd} : \Sigma^+ \rightarrow \Sigma$$
$$\text{hd} (x : \sigma) = x.$$

---

[6] Haskell Brooks Curry was a 20th century mathematical logician. He developed the ideas of Schönfinkel further on combinatory logic and later applied it to the foundations of mathematics.

The tail of a nonempty sequence is given by:

$$tl : \Sigma^+ \rightarrow \Sigma^*$$
$$tl\ (x : \sigma) = \sigma.$$

Clearly, for a nonempty sequence σ it follows that:

$$hd\ (\sigma) : tl\ (\sigma)\ = \sigma.$$

The function *len* gives the length of a sequence (i.e., the number of elements in the sequence), and is a monoid homomorphism from $(\Sigma^*, \cap, \Lambda)$ to $(\mathbb{N}, +, 0)$. The length of the empty sequence is clearly 0: i.e., $len(\Lambda) = 0$. The length of a sequence is also denoted by $|\sigma|$ or $\#\sigma$.

The elements of a sequence are given by the function *elems*. This is a monoid homomorphism from $(\Sigma^*, \cap, \Lambda)$ to $(\mathbb{P}\Sigma, \cup, \varnothing)$.

$$elems : \Sigma^* \rightarrow \mathbb{P}\Sigma$$
$$elems\ (\Lambda) = \varnothing$$
$$elems\ (x : \sigma) = \{x\} \cup elems\ (\sigma)$$

The elements of the empty sequence is the empty set $\varnothing$. The *elems* homomorphism loses information (e.g., the number of occurrences of each element in the sequence and the order in which the elements appear in the sequence). There is another operator (*items*) that determines the number of occurrences of each element in the sequence. The operator *items* generates a bag of elements from the sequence:

$$items : \Sigma^* \rightarrow (\Sigma \rightarrow \mathbb{N}_1).$$

The concatenation of two sequences is defined formally as:

$$-^\cap- : \Sigma^* \times \Sigma^* \rightarrow \Sigma^*$$

$$\Lambda^\cap \sigma = \sigma$$
$$(x : \sigma)^\cap \tau = x : (\sigma^\cap \tau)$$

The jth element in a sequence σ is given by $\sigma[i]$ where $1 \le i \le len(\sigma)$. The reversal of a sequence σ is given by $rev\ \sigma$.

6.7 Indexed Structures

An indexed monoid $(X \rightarrow M', \circledast, \theta)^7$ is created from an underlying base monoid $(M, *, u)$ and an index set X. It is defined as follows:

$$\circledast : (X \rightarrow M') \times (X \rightarrow M') \rightarrow (X \rightarrow M')$$

$$\mu \circledast \theta \qquad\qquad \underset{\Delta}{=} \mu$$

$$\mu \circledast (\{x \mapsto m\} \sqcup v) \underset{\Delta}{=} (\mu \sqcup \{x \mapsto m\}) \circledast v \qquad x \notin \mu$$

$$(\mu \dagger \{x \mapsto \mu\,(x)\,*m\}) \circledast v \qquad x \in \mu \wedge \mu\,(x)\,*m \neq u$$

$$\mu \circledast v \qquad x \in \mu \wedge \mu\,(x)\,*m = u$$

Indexing generates a higher monoid from the underlying base monoid, and this allows a chain (tower) of monoids to be built, with each new monoid built from the one directly underneath it in the chain. The power of the indexed monoid theorem is that it allows new structures to be built from existing structures, and the indexed structures inherit the properties of the underlying base structure.

A simple example of an indexed structure is a bag of elements from the set X. The indexed monoid is $(X \rightarrow \mathbb{N}_1, \oplus, \theta)$, and the underlying base monoid is $(\mathbb{N}, +, 0)$. Other indexed structures have also been considered in the Irish school of VDM.

6.8 Specifications and Proofs

Consider the specification of a simple dictionary in [But:00] where a dictionary is considered to be a set of words, and the dictionary is initially empty. There is an operation to insert a word into the dictionary, an operation to lookup a word in the dictionary, and an operation to delete a word from the dictionary.

$$w \in Word$$
$$\delta : Dict = \mathbb{P}\,Word$$
$$\delta_0 : Dict$$
$$\delta_0 \underset{\Delta}{=} \varnothing$$

The invariant is a condition (predicate expression) that is always true of the specification. The operations are required to preserve the invariant whenever the preconditions for the operations are true, and the initial system is required to satisfy the invariant. This gives rise to various proof obligations for the system.

---

[7] Recall $M' = \triangleleft \llbracket u \rrbracket\ M$

The simple dictionary above is too simple for an invariant, but in order to illustrate the concepts involved, an artificial invariant that stipulates that all words in the dictionary are British English is considered part of the system.

$$\text{isBritEng} : Word \rightarrow \mathbf{B}$$

$$\text{inv-}Dict : Dict \rightarrow \mathbf{B}$$
$$\text{inv-}Dict \; \delta \underline{\Delta} \quad \forall \; [\![\text{isBritEng}]\!] \; \delta$$

The signature of $\forall$ is $(X \rightarrow \mathbf{B}) \rightarrow \mathbb{P} X \rightarrow \mathbf{B}$, and it is being used slightly differently from the predicate calculus. There is a proof obligation to show that the initial state of the dictionary (i.e., δ_0) satisfies the invariant. That is, it is required to show that inv-$Dict \; \delta_0 = \text{TRUE}$. However, this is clearly true since the dictionary is empty in the initial state.

The first operation considered is the operation to insert a word into the dictionary. The precondition to the operation is that the word is not currently in the dictionary and that the word is British English.

$$Ins : Word \rightarrow Dict \rightarrow Dict$$
$$Ins \; [\![w]\!] \; \delta \underline{\Delta} \quad \delta \cup \{w\}$$

$$\text{pre-}Ins : Word \rightarrow Dict \rightarrow \mathbf{B}$$
$$\text{pre-}Ins \; [\![w]\!] \; \delta \underline{\Delta} \; \text{isBritEng} \; (w) \wedge w \notin \delta$$

There is a proof obligation associated with the Ins operation. It states that if the invariant is true, and the precondition for the Ins operation is true, then the invariant is true following the Ins operation.

$$\text{inv-}Dict \; \delta \wedge \text{pre-}Ins \; [\![w]\!] \; \delta \Rightarrow \text{inv-}Dict \; (Ins \; [\![w]\!] \delta)$$

COMMENT
One key difference between the Irish school of VDM and other methods such as standard VDM or Z is that postconditions are not employed in VDM$^{\clubsuit}$. Instead, the operation is explicitly constructed.

THEOREM
$\text{inv-}Dict \; \delta \wedge \text{pre-}Ins [\![w]\!] \; \delta \Rightarrow \text{inv-}Dict \; (Ins \; [\![w]\!] \; \delta)$

PROOF
$\text{inv-}Dict \; \delta \wedge \text{pre-}Ins \; [\![w]\!] \; \delta$
$\Rightarrow \forall \; [\![\text{isBritEng}]\!] \; \delta \wedge \text{isBritEng} \; (w) \wedge w \notin \delta$
$\Rightarrow (\forall \; w_d \in \delta \; \text{isBritEng} \; (w_d)) \wedge \text{isBritEng} \; (w) \wedge w \notin \delta$
$\Rightarrow (\forall \; w_d \in \delta \cup \{w\} \; \text{isBritEng} \; (w_d))$
$\Rightarrow \forall \; [\![\text{isBritEng}]\!] \; (\delta \cup \{w\})$
$\Rightarrow \text{inv-}Dict \; (\text{Ins} \; [\![w]\!] \; \delta)$

The next operation considered is a word lookup operation, and this operation returns true if the word is present in the dictionary and false otherwise. It is given by:

$$Lkp : Word \rightarrow Dict \rightarrow \mathbf{B}$$
$$Lkp \; [\![w]\!] \; \delta \underline{\Delta} \; \chi \; [\![w]\!] \; \delta$$

The final operation considered is a word removal operation. This operation removes a particular word from the dictionary and is given by:

$$Rem : Word \rightarrow Dict \rightarrow Dict$$
$$Rem \; [\![w]\!] \; \delta \underline{\Delta} \; \vartriangleleft [\![w]\!] \; \delta^8$$

There is a proof obligation associated with the *Rem* operation. It states that if the invariant is true, and the precondition for the *Rem* operation is true, then the invariant is true following the Rem operation.

$$\text{inv-}Dict \; \delta \; \wedge \text{pre-}Rem \; [\![w]\!] \; \delta \Rightarrow \text{inv-}Dict \; (Rem \; [\![w]\!] \; \delta)$$

6.9 Refinement

A specification in the Irish school of VDM involves defining the state of the system and then specifying various operations. The formal specification is implemented by a programmer, and mathematical proof is employed to provide confidence that the program meets its specification. VDM* employs many constructs that are not part of conventional programming languages, and hence, there is a need to write an intermediate specification that is between the original specification and the eventual program code. The intermediate specification needs to be correct with respect to the specification, and the program needs to be correct with respect to the intermediate specification. This requires mathematical proof.

The representation of an abstract data type like a set by a sequence is termed data reification, and data reification is concerned with the process of transforming an abstract data type into a concrete data type. The abstract and concrete data types are related by the retrieve function, and the retrieve function maps the concrete data type to the abstract data type. There are typically several possible concrete data types for a particular abstract data type (refinement is a relation), whereas there is one abstract data type for a concrete data type (i.e., retrieval is a function). For example, sets are often reified to unique sequences;

---

[8] Notation is often abused and this should strictly be written as $\vartriangleleft [\![w\}\!] \; \delta$.

however, more than one unique sequence can represent a set, whereas a unique sequence represents exactly one set.

The operations defined on the concrete data type need to be related to the operations defined on the abstract data type. The commuting diagram property is required to hold; i.e., for an operation □ on the concrete data type to correctly model the operation ⊙ on the abstract data type the following diagram must commute, and the commuting diagram property requires proof.

$retr(\sigma) \odot retr(\tau)$

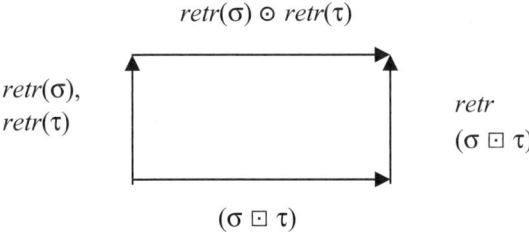

$retr(\sigma),$
$retr(\tau)$

$retr$

$(\sigma \boxdot \tau)$

$(\sigma \boxdot \tau)$

Fig. 6.4. Commuting Diagram Property

That is, it is required to prove that:

$$ret\,(\sigma \boxdot \tau) = (ret\,\sigma) \odot (ret\,\tau).$$

It needs to be proved that the initial states correspond to one another, and that each operation in the concrete state is correct with respect to the operation in the abstract state, and also that it is applicable (i.e., whenever the abstract operation may be performed the concrete operation may also be performed).

The process of refinement of the dictionary from a set to a sequence of words is considered. This involves defining the concrete state and the operations on the state, and proving that the refinement is valid. The retrieve function derives the abstract state from the concrete state, and is given by the *elems* operator for the set to sequence refinement of the dictionary. The following is adapted from [But:00]:

$$\sigma \in DSeq = Word^*$$
$$\sigma_0 : Dseq$$
$$\sigma_0 \;\underline{\Delta}\; \Lambda$$

$$inv\text{-}Dseq \;\underline{\Delta}\; \forall \,\llbracket isBritEng \rrbracket\, \sigma$$

$$retr\text{-}Dict : DSeq \rightarrow Dict$$
$$retr\text{-}Dict\;\sigma \,\underline{\Delta}\; elems\;\sigma$$

Here, $\forall$ has signature $(X \rightarrow \mathbf{B}) \rightarrow X^* \rightarrow \mathbf{B}$.

The first operation considered on the concrete state is the operation to insert a word into the dictionary.

$$Ins_1 : Word \rightarrow DSeq \rightarrow DSeq$$
$$Ins_1 \; \llbracket w \rrbracket \; \sigma \; \underline{\Delta} \; w : \sigma$$

$$\text{pre-}Ins_1 : Word \rightarrow DSeq \rightarrow \mathbf{B}$$
$$\text{pre-}Ins_1 \; \llbracket w \rrbracket \; \sigma \; \underline{\Delta} \; \text{isBritEng} \,(w) \wedge w \notin \text{elems} \,(\sigma\,)$$

There is a proof obligation associated with the Ins_1 operation.

$$\text{inv-}DSeq \; \sigma \; \wedge \text{pre-}Ins_1 \; \llbracket w \rrbracket \; \sigma \Rightarrow \text{inv-}DSeq(Ins_1 \; \llbracket w \rrbracket \; \sigma\,)$$

The proof is similar to that considered earlier on the abstract state. Next, we show that Ins_1 is a valid refinement of Ins. This requires that the commuting diagram property holds:

$$\text{pre-}Ins_1 \; \llbracket w \rrbracket \; \sigma \Rightarrow \text{retr-}Dict(Ins_1 \; \llbracket w \rrbracket \; \sigma) = Ins \; \llbracket w \rrbracket \; (\text{retr-}Dict \; \sigma)$$

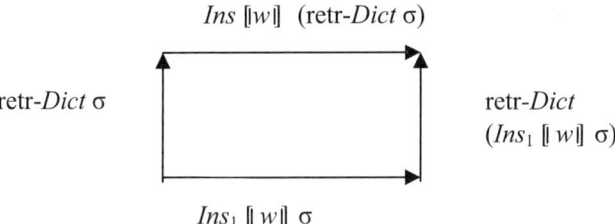

$Ins \; \llbracket w \rrbracket \; (\text{retr-}Dict \; \sigma)$

retr-*Dict* σ

retr-*Dict*
$(Ins_1 \; \llbracket w \rrbracket \; \sigma)$

$Ins_1 \; \llbracket w \rrbracket \; \sigma$

Fig. 6.5. Commuting Diagram for Dictionary Refinement

PROOF
pre-$Ins_1 \; \llbracket w \rrbracket \; \sigma$
$\Rightarrow \text{isBritEng} \,(w) \wedge w \notin \text{elems} \,(\sigma\,)$

retr-$Dict(Ins_1 \; \llbracket w \rrbracket \; \sigma)$
$= \text{retr-}Dict \; (w : \sigma)$
$= elems \; (w : \sigma)$
$= \{w\} \cup elems \; (\sigma)$
$= \{w\} \cup \text{retr-}Dict \; (\sigma)$
$= Ins \; \llbracket w \rrbracket \; (\text{retr-}Dict \; \sigma)$

There are other operations for the concrete representation of the dictionary and these are discussed in [But:00].

6.10 Summary

The Irish School of VDM is a variant of standard VDM, and is characterized by its constructive approach, classical mathematical style, and its terse notation. The method combines the *what* and *how* of formal methods in that its terse specification style stipulates in concise form *what* the system should do; and furthermore the fact that its specifications are constructive (or functional) means that that the *how* is included with the *what*.

VDM* follows a similar development methodology as in standard VDM and is a model-oriented approach. The initial specification is presented, with initial state and operations defined. The operations are presented with preconditions; and the operation is functionally constructed. Each operation has an associated proof obligation; if the precondition for the operation is true and the operation is performed, then the system invariant remains true after the operation.

The school has been influenced by the work of Polya and Lakatos. Polya has recommended problem solving by first tackling easier subproblems, whereas Lakatos adopted a heuristic approach to mathematical discovery based on proposing theorems and discovering hidden lemmas.

There is a rich operator calculus in the Irish school of VDM, and new operators and structures that are useful for specification and proof are sought. A special emphasis is placed on the identification of useful structures and their morphisms that provide compact specifications and proof.

Partial functions are employed, and care is taken to ensure that the function is defined and will terminate prior to function application. The logic of partial functions (LPFs) is avoided and care is taken to ensure that the recursive argument is strictly decreasing in each recursive invocation. The $\perp$ symbol is typically used in the Irish school to represent *undefined or unavailable* or *do not care*. Academic and industrial projects have been conducted using VDM*, but at this stage tool support is limited.

The formal methods group at Trinity College, Dublin (www.cs.tcd.ie/fmg) is active in promoting the philosophy and method of the Irish school of VDM. Further information on VDM* is available in [Mac:90, But:99].

7
Dijkstra and Hoare

7.1 Introduction

Edsger W. Dijkstra and C. A. R. Hoare are two famous names in computer science, and both have received numerous awards for their contribution to the discipline. Their work has provided a scientific basis for computer software.[1] Dijkstra was born in Rotterdam in Holland and studied mathematics and physics at the University of Leyden. He obtained a PhD in computer science from the University of Amsterdam in 1959. He decided not to become a theoretical physicist, as he believed that programming offered a greater intellectual challenge than theoretical physics.

He commenced his programming career at the Mathematics Centre in Amsterdam in the early 1950s, and discovered the shortest path algorithm in the mid-1950s. He contributed to the definition of Algol 60, and it became a standard in 1962. Dijkstra then designed and coded the first Algol 60 compiler. He became a professor of mathematics at Eindhoven University, Holland, in the early 1960s, and became a Burroughs research fellow in 1973. He received the Turing award in 1972 and took a position at the University of Texas in Austin in 1984. He died of cancer in 2002.

Dijkstra made many contributions to computer science, including contributions to language development, operating systems, formal program development, and to the vocabulary of computer science. Some of his achievements are summarized in the table below:

Area	Description
Go to statement	Dijkstra argued against the use of the go to statement in programming. This eventually led to the abolition of its use in programming languages.

---

[1] However, it remains to be seen if the theoretical foundations developed by Dijkstra and Hoare can be made usable to practitioners in mainstream software engineering.

Graph algorithms	Dijkstra developed various efficient graph algorithms to determine the *shortest* or *longest* paths from a vertex u to vertex v in a graph.
Operating systems	Dijkstra discovered that operating systems can be built as synchronized sequential processes. He introduced ideas such as semaphores and deadly embrace.
Algol 60	Dijkstra contributed to the definition of the language and designed and coded the first Algol 60 compiler.
Formal program development (Guarded commands and predicate transformers)	Dijkstra introduced guarded commands and predicate transformers as a means of defining the semantics of a programming language. He showed how weakest preconditions can be used as a calculus (*wp*-calculus) to develop reliable programs. This led to a science of programming using mathematical logic as a methodology for formal program construction. He applied mathematical proof techniques to programming, and his approach involved the development of programs from mathematical axioms.

Table 7.1. Dijkstra's Achievements.

The approach to building high-quality software in the 1960s was to try out the program on test cases until they were debugged. John McCarthy argued at the IFIP congress in 1962 that the focus should instead be to prove that the programs have the desired properties rather than testing the program *ad nauseum*. The NATO conference on software engineering in 1968 highlighted the extent of the problems that existed with software, and the term *software crisis* was coined to describe this. The problems included cost and schedule overruns and problems with software reliability.

Dijkstra advocated simplicity, precision, and mathematical integrity in his formal approach to program development using the weakest precondition calculus. He insisted that programs should be composed correctly using mathematical techniques, and not debugged into correctness. He considered testing to be an inappropriate means of building quality into software, and his statement on software testing is well known:

Testing a program shows that it contains errors never that it is correct.[2]

Dijkstra corresponded with other academics through an informal distribution network known as the EWD series. These contain his various personal papers including trip reports and technical papers.

Charles Anthony Richard (Tony) Hoare studied philosophy (including Latin and Greek) at Oxford University. He studied Russian at the Royal Navy during his National Service in the late 1950s. He then studied statistics and went to Moscow University as a graduate student to study machine translation of languages and probability theory. He discovered the well-known sorting algorithm *Quicksort* while investigating efficient ways to look up words in a dictionary. He returned to England in 1960 and worked as a programmer for Elliot Brothers (a company that manufactured scientific computers). He led a team to produce the first commercial compiler for Algol 60 and the project was very successful. He then led a team to implement an operating system, and the project was a disaster. He managed a recovery from the disaster and then moved to the research division of the company. He took a position at Queens University in Belfast, Ireland, in 1968, and his research goals included examining techniques to assist with the implementation of operating systems, especially to see if advances in programming methodologies could assist with the problems of concurrency. He also published material on the use of assertions to prove program correctness. He moved to Oxford University in 1977 following the death of Christopher Strachey (well known for his work in denotational semantics), and built up the programming research group. This group has produced the Z specification language and CSP. He received the Turing Award in 1980. Following his retirement from Oxford he took up a position as senior researcher at Microsoft Research in the United Kingdom.

Hoare has made many contributions to computer science, and these include the Quicksort algorithm, the axiomatic approach to program semantics, and programming constructs for concurrency. These are summarized in the table below:

Area	Description
Quicksort	Quicksort is a highly efficient sorting algorithm and is used often by programmers.
Axiomatic semantics	Hoare defined a small programming language in terms of axioms and logical inference rules. He then showed how this could be applied to prove partial correctness of programs.

---

[2] I see software testing as an essential part of the software process, and various types of testing are described in [ORg:02]. Modern software testing is quite rigorous and can provide a high degree of confidence that the software is fit for use. It cannot, of course, build quality in; rather, it can give confidence that quality has been built in. The analysis of the defects identified during testing may be useful in improving the software development process.

Communicating se-quential processes (CSP)	CSP is a mathematical approach to the study of communication and concurrency. It is applicable to the specification and design of computer systems that continuously interact with their environment.

Table 7.2. Hoare's Achievements.

Hoare was responsible for producing the first commercial compiler for Algol 60 during his period working at Elliot Brothers. He later remarked later:

Algol 60 was a great achievement in that it was a significant advance over most of its successors.

Hoare has made fundamental contributions to programming languages and his ACM lecture on the "Emperors Old Clothes"[3] is well known. He stresses the importance of communicating ideas (as well as having ideas), and enjoys writing (and rewriting). A paper by Hoare may include eight or nine rewrites[4] before publication.

7.2 Calculus of Weakest Preconditions

The weakest precondition calculus was developed by Dijkstra [Dij:76] and has been applied to the formal development of programs. The material presented here is based on [Gri:81], and a programming notation is introduced and defined in terms of the weakest precondition. The weakest precondition $wp(S,R)$ is a predicate that describes a set of states. It is a function with two arguments (a command and a predicate that describes the desired result after the execution of the command). It is defined as follows:

DEFINITION (WEAKEST PRECONDITION)
The predicate $wp(S,R)$ represents the set of all states such that, if execution of S commences in any one of them, then it is guaranteed to terminate in a state satisfying R.

---

[3] Hoare argues for simplicity in language design and in software design and argues against languages such as Algol 68 and ADA which attempted to be all things to all people. There is a well-known quote in which he states "There are two ways of constructing a software design. One way is to make it so simple that there are obviously no deficiencies, and the other way is to make it so complex that there are no obvious deficiencies."

[4] Parnas has also noted the importance of rewriting and states that one of the people who influenced him in his early years stressed the importance of using waste paper baskets to dispose of the current draft paper and start again.

Let S be the assignment command $i := i+5$ and let R be $i \leq 3$ then

$$wp(i := i+5;\ i \leq 3) = (\ i \leq -2).$$

The weakest precondition $wp(S,T)$ represents the set of all states such that if execution of S commences in any one of them, then it is guaranteed to terminate:

$$wp(i := i+5;\ T) = T.$$

The weakest precondition $wp(S,R)$ is a precondition of S with respect to R and it is also the weakest such precondition. Given another precondition P of S with respect to R, then $P \Rightarrow wp(S,R)$.

For a fixed command S then $wp(S,R)$ can be written as a function of one argument $wp_S(R)$, and the function wp_S transforms the predicate R to another predicate $wp_S(R)$; i.e., the function wp_S acts as a *predicate transformer*.

An imperative program may be regarded as a predicate transformer. This is since a predicate P characterizes the set of states in which the predicate P is true, and an imperative program may be regarded as a binary relation on states, which may be extended to a function F, leading to the Hoare triple $P\{F\}Q$. That is, the program F acts as a predicate transformer. The predicate P may be regarded as an input assertion, i.e., a Boolean expression which must be true before the program F is executed. The Boolean expression Q is the output assertion, and is true if the program F terminates, having commenced in a state satisfying P.

Properties of WP

The weakest precondition $wp(S,R)$ has several well-behaved properties such as:

- **Law of the excluded miracle**
  $$wp(S,\ F) = F$$

This describes the set of states such that if execution commences in one of them, then it is guaranteed to terminate in a state satisfying false. However, no state ever satisfies false, and therefore $wp(S,\ F) = F$. The name of this law derives from the fact that it would be a miracle if execution could terminate in no state.

- **Distributivity of Conjunction**
  $$wp(S,\ Q) \wedge wp(S,\ R) = wp(S,\ Q \wedge R)$$

This property stipulates that the set of states such that if execution commences in one of them, then it is guaranteed to terminate in a state satisfying $Q \wedge R$ is precisely the set of states such that if execution commences in one of them then execution terminates with both Q and R satisfied.

- **Law of Monotonicity**

$$Q \Rightarrow R \text{ then } wp(S, Q) \Rightarrow wp(S, R)$$

This property states that if a postcondition Q is stronger than a post-condition R, then the weakest precondition of S with respect to Q is stronger than the weakest precondition of S with respect to R.

- **Distributivity of Disjunction**

$$wp(S, Q) \vee wp(S, R) \Rightarrow wp(S, Q \vee R)$$

This property states that the set of states corresponding to the weakest precondition of S with respect to Q or the set of states corresponding to the weakest precondition of S with respect to R is stronger than the weakest precondition of S with respect to $Q \vee R$. Equality holds for distributivity of disjunction only when the execution of the command is deterministic.

WP of Commands

The weakest precondition can be used to provide the definition of commands in a programming language. The commands considered here are as in [Gri:81].

- **Skip Command**

$$wp(skip, R) = R$$

The *skip* command does nothing and is used to explicitly say that nothing should be done. The predicate transformer wp_{skip} is the identity function.

- **Abort Command**

$$wp(abort, R) = F$$

The *abort* command is executed in a state satisfying false (i.e., no state). This command should never be executed. If program execution reaches a point where *abort* is to be executed then the program is in error and abortion is called for.

- **Sequential Composition**

$$wp(S_1;S_2, R) = wp(S_1, wp(S_2,R))$$

The sequential composition command composes two commands S_1 and S_2 by first executing S_1 and then executing S_2. Sequential composition is expressed by $S_1;S_2$. Sequential composition is associative:

$$wp(S_1;(S_2; S_3),R) = wp((S_1;S_2); S_3, R).$$

• **Simple Assignment Command**

$$wp(x := e, R) = dom(e) \textbf{ cand } R^x_e$$

The execution of the *assignment* command consists of evaluating the value of the expression e and storing its value in the variable x. However, the command may be executed only in a state where e may be evaluated. The expression R^x_e denotes the expression obtained by substituting e for all free occurrences of x in R. For example:

$$(x + y > 2)^x_v = v + y > 2.$$

The **cand** operator is used to deal with undefined values and was discussed in Chapter 3. It is a noncommutative operator and its truth table is defined in [Gri:81]. The expression a **cand** b is equivalent to:

$$a \textbf{ cand } b \cong \textbf{ if } a \textbf{ then } b \textbf{ else } F.$$

The explanation of the definition of the weakest precondition of the assignment statement $wp(x := e, R)$ is that R will be true after execution if and only if the predicate R with the value of be true x replaced by e is true before execution (since x will contain the value of e after execution). Often, the domain predicate $dom(e)$ that describes the set of states that e may be evaluated in is omitted as assignments are usually written in a context in which the expressions are defined:

$$wp(x := e, R) = R^x_e.$$

The simple assignment can be extended to a multiple assignment to simple variables. The assignment is of the form $x_1, x_2, ..., x_n := e_1, e_2, ..., e_n$ and is described in [Gri:81].

• **Assignment to Array Element Command**

$$wp(b[i] := e, R) = inrange(b, i) \textbf{ cand } dom(e) \textbf{ cand } R^b_{(b; i : e)}$$

The execution of the *assignment to an array element* command consists of evaluating the expression e and storing its value in the array element subscripted by i. The $inrange(b, i)$ and $dom(e)$ are usually omitted in practice as assignments are usually written in a context in which the expressions are defined and the subscripts are in range. Therefore, the weakest precondition is given by:

$$wp(b[i] := e, R) = R^b_{(b; i : e)}$$

The notation $(b; i : e)$ denotes an array identical to array b except that the array element subscripted by i contains the value e. The explanation of the definition of the weakest precondition of the assignment statement to an array element $(wp(b[i] := e, R)$ is that R will be true after execution if and only if the value of b replaced by $(b;i:e)$ is true before execution (since b will become $(b;i:e)$ after execution).

• Alternate Command

$$wp(IF, R) = dom(B_1 \lor B_2 \lor \ldots \lor B_n) \land (B_1 \lor B_2 \lor \ldots \lor B_n)$$
$$\land (B_1 \Rightarrow wp(S_1, R)) \land (B_2 \Rightarrow wp(S_2, R)) \land \ldots \land (B_n \Rightarrow wp(S_n, R))$$

The *alternate* command is the familiar **if** statement of programming languages. The general form of the alternate command is:

> **If** $B_1 \rightarrow S_1$
> ☐ $B_2 \rightarrow S_2$
> …
> ☐ $B_n \rightarrow S_n$
> **fi**

Each $B_i \rightarrow S_i$ is a guarded command (S_i is any command). The guards must be well defined in the state where execution begins, and at least one of the guards must be true or execution aborts. If at least one guard is true, then one guarded command $B_i \rightarrow S_i$ with true guard B_i is chosen and S_i is executed. For example, in the if statement below, the statement $z := x+1$ is executed if $x > 2$, and the statement $z := x+2$ is executed if $x < 2$. For $x = 2$ either (but not both) statements are executed. This is an example of non-determinism.

> **if** $x \geq 2 \rightarrow z := x+1$
> ☐ $x \leq 2 \rightarrow z := x+2$
> **fi**

• Iterative Command

The *iterate* command is the familiar while loop statement of programming languages. The general form of the iterate command is:

> **do** $B_1 \rightarrow S_1$
> ☐ $B_2 \rightarrow S_2$
> …
> ☐ $B_n \rightarrow S_n$
> **od**

The meaning of the iterate command is that a guard B_i is chosen that is true and the corresponding command S_i is executed. The process is repeated until there are no more true guards. Each choice of a guard and execution of the corresponding statement is an iteration of the loop. On termination of the iteration command all of the guards are false. The meaning of the DO command $wp(DO, R)$ is the set of states in which execution of DO terminates in a bounded number of iterations with R true.

$$wp(DO, R) = (\exists k : 0 \le k : H_k(R))$$

where $H_k(R)$ is defined as:

$$H_k(R) = H_0(R) \vee wp(\text{IF}, H_{k-1}(R))$$

A more detailed explanation of loops is in [Gri:81]. The definition of procedure call may also be given in weakest preconditions.

Formal Program Development with WP

The use of weakest preconditions for formal program development is described in [Gri:81]. The approach is a radical departure from current software engineering and involves formal proofs of correctness.[5] A program P is correct with respect to a precondition Q and a postcondition R if $\{Q\}P\{R\}$. The idea is that the program and its proof should be developed together. The proof involves weakest preconditions and uses the formal definition of the programming constructs (e.g., assignment and iteration) discussed earlier.

Programming is viewed as a goal-oriented activity in that the desired result (i.e., the postcondition R) plays a more important role in the development of the program than the precondition Q. Programming is employed to solve a problem, and the problem needs to be clearly stated with precise pre- and postconditions.

The example of a program[6] P to determine the maximum of two integers x and y is discussed in [Gri:81]. A program P is required that satisfies:

$$\{T\}P\{R: z = max(x,y)\}$$

The postcondition R is then refined by replacing max with its definition:

$$\{R: z \ge x \wedge z \ge y \wedge (z = x \vee z = y).\}$$

The next step is to identify a command that could be executed in order to establish the postcondition R. One possibility is $z := x$ and the conditions under which this assignment establishes R is given by:

$$\begin{aligned} wp(\text{“}z := x\text{”}, R) \quad &= x \ge x \wedge x \ge y \wedge (x = x \vee x = y) \\ &= x \ge y \end{aligned}$$

---

[5] I consider Dijkstra's weakest preconditions to be a nice elegant theoretical approach that has limited direct application to mainstream software engineering. I find the approach slightly cumbersome but this may be due to my own failings.

[6] I see this as a toy example and note that the formal methods community is extremely good at producing toy examples. However, what is required is the development of a suite of practical industrial examples to demonstrate the usefulness and applicability of the various theoretical approaches.

Another possibility is $z := y$ and the conditions under which this assignment establishes R is given by:

$$\text{wp}(\text{“}z := y\text{”}, R) \quad = y \geq x$$

The desired program is then given by:

if $x \geq y \rightarrow z := x$
$\square$ $y \geq x \rightarrow z := y$
fi

There are many more examples of formal program development in [Gri:81]. The next section considers early work done by C. A. R Hoare on the axiomatic semantics of programming languages. The axiomatic semantics of programming languages give a precise meaning to the various constructs in the language.

7.3 Axiomatic Definition of Programming Languages

An assertion is a property of the program's objects: e.g., the assertion $(x - y > 5)$ is an assertion that may or may not be satisfied by a state of the program during execution. An assertion is essentially a Boolean expression. For example, a state in which the values of the variables x and y are 7 and 1 respectively satisfies the assertion; whereas a state in which x and y have values 4 and 2 respectively does not.

The first article on program proving using techniques based on assertions was by Floyd in 1967 [Fly:67]. The paper was concerned with assigning meaning to programs and also introduced the idea of a loop invariant. Floyd's approach was based on programs expressed by flowcharts, and an assertion was attached to the edge of the flowchart. The meaning was that the assertion would be true during execution of the corresponding program whenever execution reached that edge. For a loop, Floyd placed an assertion P on a fixed position of the cycle and proved that if execution commenced at the fixed position with P true then and reached the fixed position again, then P would still be true.

Hoare refined and improved upon Floyd's work in 1969 [Hor:69], and he proposed a logical system for proving properties of program fragments. The well-formed formulae of the logical system are of the form:

$$P \{a\} Q$$

where P is the precondition; a is the program fragment; and Q is the postcondition. The precondition P is a predicate (or input assertion), and the postcondition R is a predicate (output assertion). The braces separate the assertions from the program fragment. The well-formed formula P $\{a\}Q$ is itself a predicate that is either true or false. This notation expresses the partial correctness of a with respect to P and Q where partial correctness and total correctness are defined as follows:

DEFINITION (PARTIAL CORRECTNESS)
A program fragment a is partially correct for precondition P and postcondition Q if and only if whenever a is executed in any state in which P is satisfied and this execution terminates, then the resulting state satisfies Q.

The proof of partial correctness requires proof that the postcondition Q is satisfied if the program terminates. Partial correctness is a useless property unless termination is proved, as any nonterminating program is partially correct with respect to any specification.

DEFINITION (TOTAL CORRECTNESS)
A program fragment a is totally correct for precondition P and postcondition Q if and only if whenever a is executed in any state in which P is satisfied the execution terminates and the resulting state satisfies Q.

The proof of total correctness requires proof that the postcondition Q is satisfied and that the program terminates. Total correctness is expressed by $\{P\}$ a $\{Q\}$. The calculus of weakest preconditions developed by Dijkstra discussed in the previous section is based on total correctness, whereas Hoare's approach is based on partial correctness.

Hoare's axiomatic theory of programming languages consists of axioms and rules of inference to derive certain pre-post formulae. The meaning of several constructs in programming languages is presented here in terms of pre-post semantics.

• Skip
The meaning of the skip command is:

$$P \{skip\} P.$$

The skip command does nothing and this instruction guarantees that whatever condition is true on entry to the command is true on exit from the command.

• Assignment
The meaning of the assignment statement is given by the axiom:

$$P^x_e \{x:=e\}P.$$

The notation P^x_e has been discussed previously and denotes the expression obtained by substituting e for all free occurrences of x in P. The meaning of the assignment statement is that P will be true after execution if and only if the predicate P^x_e with the value of x replaced by e in P is true before execution (since x will contain the value of e after execution).

• **Compound**
The meaning of the conditional command is:

$$\frac{P\ \{S_1\}Q,\ \ Q\ \{S_2\}R}{P\ \{S_1\ ;\ S_2\ \}R}$$

The execution of the **compound** statement involves the execution of S_1 followed by S_2. The correctness of the compound statement with respect to P and R is established by proving that the correctness of S_1 with respect to P and Q and the correctness of S_2 with respect to Q and R.

• **Conditional**
The meaning of the conditional command is:

$$\frac{P{\wedge}B\ \{S_1\}Q,\ \ \ \ \ \ \ P{\wedge}{\neg}B\ \{S_2\}Q}{P\{\textbf{if}\ B\ \textbf{then}\ S_1\ \textbf{else}\ S_2\}Q}$$

The execution of the **if** statement involves the execution of S_1 or S_2. The execution of S_1 takes place only when B is true, and the execution of S_2 takes place only when $\neg B$ is true. The correctness of the **if** statement with respect to P and Q is established by proving that S_1 and S_2 are correct with respect to P and Q.

However, S_1 is executed only when B is true, and therefore it is required to prove the correctness of S_1 with respect to $P{\wedge}B$ and Q, and the correctness of S_2 with respect to $P{\wedge}{\neg}B$ and Q.

• **While Loop**
The meaning of the while loop is given by:

$$\frac{P{\wedge}B\ \{S\}\ P}{P\ \{\textbf{while}\ B\ \textbf{do}\ S\}\ P{\wedge}{\neg}B}$$

The property P is termed the loop invariant as it remains true throughout the execution of the loop. The invariant is satisfied before the loop begins and each iteration of the loop preserves the invariant.

The execution of the **while loop** is such that if the truth of P is maintained by one execution of S, then it is maintained by any number of executions of S. The execution of S takes place only when B is true, and upon termination of the loop $P \land \neg B$ is true.

Loops may fail to terminate and there is therefore a need to prove termination. The loop invariant needs to be determined for formal program development.

7.4 Communicating Sequential Processes

The objectives of the process calculi [Hor:85] are to provide mathematical models which provide insight into the diverse issues involved in the specification, design, and implementation of computer systems which continuously act and interact with their environment. These systems may be decomposed into subsystems which interact with each other and their environment. The basic building block is the *process*, which is a mathematical abstraction of the interactions between a system and its environment.

A process which lasts indefinitely may be specified recursively. Processes may be assembled into systems, execute concurrently, or communicate with each other. Process communication may be synchronized, and generally takes the form of a process outputting a message simultaneously to another process inputting a message. Resources may be shared among several processes. Process calculi enrich the understanding of communication and concurrency, and an elegant formalism such as CSP [Hor:85] obeys a rich collection of mathematical laws.

The expression $(a \rightarrow P)$ in CSP describes a process which first engages in event a, and then behaves as process P. For example, a vending machine (as adapted from [Hor:85]) that serves one customer before breaking is given by:

$$(coin \rightarrow (choc \rightarrow STOP)).$$

A recursive definition is written as $(\mu X):A \bullet F(X)$, where A is the alphabet of the process. The behavior of a simple chocolate vending machine is given by the following recursive definition:

$$VMS = \mu X:\{coin, choc\}.(coin \rightarrow (choc \rightarrow X)).$$

The simple vending machine has an alphabet of two symbols, namely, *coin* and *choc*, and the behavior of the machine is such that when a coin is entered into the machine, a chocolate is then provided. This machine repeatedly serves chocolate in response to a coin.

It is reasonable to expect the behavior of a process to be influenced by interaction with its environment. A vending machine may also provide tea and coffee as well as chocolate, and it is the customer's choice as to which product is selected at the machine. The choice operation is used to express the choice of two distinct events x and y. The subsequent behavior is described by P if the first event was x, and otherwise Q:

$$(x \rightarrow P \mid y \rightarrow Q)$$

The following machine serves either chocolate or tea on each transaction:

$$VM = \mu X:(coin \rightarrow (choc \rightarrow X \mid tea \rightarrow X)).$$

The definition of choice can be extended to more than two alternatives:

$$(x \rightarrow P \mid y \rightarrow Q \mid ... \mid z \rightarrow R).$$

The recursive definition above allows the definition of a single process as the solution of a single equation. This may be generalized to mutual recursion involving the solution of sets of simultaneous equations in more than one unknown.

No.	Property	Description
1.	$(x \rightarrow P \mid y \rightarrow Q) = (y \rightarrow Q \mid x \rightarrow P)$	Choice operator is commutative.
2.	$(x \rightarrow P) \neq STOP$	A process that can do something is not the same as one that does nothing.
3.	$(x \rightarrow P) = (y \rightarrow Q) \equiv x = y \land P = Q$	Equality.
4.	$(x{:}A \rightarrow P(x))=(y{:}B \rightarrow Q(y)) \equiv (A = B \land \forall x \in A \bullet P(x) = Q(x))$	General choice equality
5.	$F(X)$ a guarded expression then $(Y=F(Y) \equiv (Y=\mu X.F(X))$	Every properly guarded recursive equation has only one solution.
6.	$\mu X.F(X) = F(\mu X.F(X))$	$\mu X.F(X)$ is the solution.

Table 7.3. Properties of Processes

The behavior of a process can be recorded as a trace of the sequences of actions that it has engaged. This allows a process to be specified prior to implementation by describing the properties of its traces. A trace of the behavior of a process is a finite sequence of symbols recording the events that the process has engaged up to a particular time instance. Suppose events $x,y,$ and z have occurred. Then this is represented by the trace:

$$\langle x,y,z \rangle$$

Initially, the process will have engaged in no events and this is represented by the empty trace:

$$\langle \rangle$$

There are various operations on traces such as concatenation, restriction of a trace to elements of a set, the head and tail of a trace, the star operator, and the interleaving of two traces. The concatenation of two traces s and t is denoted by $s \wedge t$. For example, the concatenation of $\langle x,y \rangle$ and $\langle z \rangle$ is given by $\langle x,y,z \rangle$. The restriction operator is employed on traces to restrict the symbols in the trace t to elements of a particular set A. It is denoted by $t \upharpoonright$ A and the restriction of $\langle x,z,x \rangle$ to $\{x\}$ is given by $\langle x,z,x \rangle \upharpoonright \{x\} = \langle x,x \rangle$. The head and tail of a trace s is given by s_0 and s' respectively. The set A^* is the set of all finite traces formed from the set A. The length of a trace t is given by $\#t$. A sequence s is an interleaving of two sequences u and t if s consists of all the elements of u and t and can be split into two subsequences to yield u and t. The mathematical definition of interleaving is recursive and given in [Hor:85].

A trace of a process records the behavior of the process up to a specific time instance. The particular trace that will be recorded is not known in advance as it depends on how the process interacts with its environment. However, the complete set of all possible traces of a process P can be determined in advance and is given by *traces*(P). There are various properties of the *traces* function:

No.	Property	Description
1.	traces(STOP) = $\{\langle\rangle\}$	There is only one trace of the STOP process namely the empty trace.
2.	$traces(x \rightarrow P) = \{\langle\rangle\} \cup \{\langle x \rangle \wedge t\} \mid t \in traces(P)\}$	A trace of $(x \rightarrow P)$ is either empty or begins with x such that its tail is a possible trace of P.
3.	$traces(x \rightarrow P \mid y \rightarrow Q) = \{t \mid t = \langle\rangle \vee (t_0 = x \wedge t' \in traces(P)) \vee (t_0 = y \wedge t' \in traces(Q))\}$	A trace of a process that offers a choice is a trace of one of the alternatives.
4.	$traces(\mu X : A.F(X)) = \cup_{n \geq 0} traces(F^n(STOP_A))$	$F^n(X) = F(F^{n-1}(X))$ and $F^0(X) = X$

Table 7.4. Properties of Traces.

The definition of the intended behavior of the process is termed its specification, and the implementation of a process can be proved to meet its specification (i.e., P **sat** S). Processes can be assembled together to form systems, and the processes interact with each other and their environment. The environment may also be described as a process, and the complete system is regarded as a process with its behavior defined in terms of the component processes.

The interaction between two processes may be regarded as events that require the simultaneous participation of both the processes. The interaction between two processes with the same event alphabet is denoted by:

$$P \| Q.$$

The interaction operator is well behaved and satisfies the following properties:

No.	Property	Description
1.	$P \parallel Q = Q \parallel P$	Commutative
2.	$(P \parallel Q) \parallel R = P \parallel (Q \parallel R)$	Associative
3.	$P \parallel STOP = STOP$	Deadlock
4.	$(x \to P) \parallel (x \to Q) = (x \to (P \parallel Q))$	Engagement
5.	$(x \to P) \parallel (y \to Q) = STOP$ $(x \neq y)$	Deadlock
6.	$(x{:}A \to P(x)) \parallel (y{:}B \to Q(y)) =$ $(z{:}A \cap B \to (P(z) \parallel Q(z)))$	Only events they both offer will be possible

Table 7.5. Properties of Interaction Operator

The $\parallel$ operator can be generalized to the case where the operands P and Q have different alphabets. When such processes are run concurrently, then events that are in both alphabets require the simultaneous participation of P and Q, whereas events that are in the alphabet of P but not in Q are of no concern to Q, and vice versa for events that are in the alphabet of Q but not in the alphabet of P. This is described in detail in [Hoa:85].

There are some tools available for CSP and these include FDR and ProBE from Formal Systems Europe in the United Kingdom. FDR has been employed to an extent in industry to check for safety and liveness properties such as deadlock and livelock. ProBE is an animator tool for CSP.

7.5 Summary

Edsger W. Dijkstra and C. A. R. Hoare have both made major contributions to computer science, and their work has provided a scientific basis for computer software. Dijkstra was born in Rotterdam, Holland and studied at the University of Leyden, and later obtained his PhD from the University of Amsterdam. He worked at the Mathematics Centre in Amsterdam and later in Eindhoven in the Netherlands. He later worked in Austin in the U.S. His many contributions to computer science include shortest path algorithms, Operating systems, Algol 60 and formal system development (using the weakest precondition calculus).

Hoare studied philosophy at Oxford in the United Kingdom, and took a position in computing with Elliot Brothers in the early 1960s. He discovered the Quicksort algorithm while studying machine translation at Moscow University. He moved to Queens University in Belfast in 1968 and dedicated himself to providing a scientific basis to programming. He moved to Oxford University in 1977 and following his retirement from Oxford he took a position with Micro-soft Research in the U.K. His many contributions to computer science include Quicksort, Axiomatic Definition of Programming Languages and Communicating Sequential Processes. Quicksort is a highly efficient sorting algorithm. The axiomatic definition of programming languages provides a logical system for proving properties of program fragments CSP provides a mathematical model that provide insight into the issues involved in the specification, design, and implementation of computer systems which continuously act and interact with their environment.

8
The Parnas Way

8.1 Introduction

David L. Parnas has been influential in the computing field and his ideas on the specification, design, implementation, maintenance and documentation of computer software are still relevant today. He has won numerous awards (including ACM best paper award in 1979; two most influential paper awards from ICSE in 1978 and 1984; the ACM SigSoft outstanding researcher award in 1998; and an honorary doctorate from the ETH in Zurich and the Catholic University of Louvain in Belgium) for his contribution to computer science. Software engineers today continue to use his ideas in their work.[1]

He studied at Carnegie Mellon University in the United States and was awarded B.S., M.S., and PhD degrees in electrical engineering by the university. He has worked in both industry and academia and his approach aims to achieve a middle way between theory and practice. His research has focused on real industrial problems that engineers face and on finding solutions to these practical problems. Several organizations such as Phillips in the Netherlands; the Naval Research Laboratory (NRL) in Washington; IBM Federal Systems Division; and the Atomic Energy Board of Canada have benefited from his talent, expertise and advice.

He advocates a solid engineering approach to the development of high-quality software and argues that software engineers[2] today do not have the right engineering education to perform their roles effectively. The role of engineers is to apply scientific principles and mathematics to design and develop useful products. The level of mathematics taught in most computer science courses is

---

[1] I have been surprised that some of Parnas's contributions (especially on information-hiding and its role in the object-oriented world) do not seem to be well known among students.

2 Parnas argues that the term engineer should be used only in its classical sense as a person who is qualified and educated in science and mathematics to design and inspect products. Parnas laments that the evolution of language that has led to a debasement of the term engineer to include various other groups who do not have the appropriate background to be considered engineers in the classical sense. However, the evolution of language is part of life, and therefore it is natural for the meaning of the term 'engineer' to evolve accordingly. It is a fact of life that the term engineer is now applied to others as well as to classical engineers.

significantly less than that taught to traditional engineers. In fact, computer science graduates generally enter the workplace with knowledge of the latest popular technologies but with only a limited knowledge of the foundations needed to be successful in producing safe and useful products. Consequently, it should not be surprising that the quality of software produced today falls below the desired standard as the current approach to software development is informal and based on intuition rather than sound engineering principles. He argues that computer scientists should be educated as engineers and provided with the right scientific and mathematical background to do their work effectively. This is discussed further later in this chapter.

8.2 Achievements

Parnas has made a strong contribution to software engineering. This includes over 200 research papers, including contributions to requirements specification, software design, software inspections, testing, tabular expressions, predicate logic, and ethics for software engineers. He has made significant contributions to industry and teaching. His reflections on software engineering are valuable and contain the insight gained over a long career.

Area	Description
Tabular expressions	Tabular expressions are mathematical tables that are employed for specifying requirements. They enable complex predicate logic expressions to be represented in a simpler form. They are used in mathematical documents.
Mathematical documentation	He advocates the use of mathematical documents for software engineering that are precise and complete. These include documents for the system requirements, system design, software requirements, module interface specification, and module internal design.
Requirements specification	His approach to requirements specification (developed with Kathryn Heninger and others) is mathematical. It involves the use of mathematical relations to specify the requirements precisely.

Software design	His contribution to software design was revolutionary. A module is characterized by its knowledge of a design decision (secret) that it hides from all others. This is known as the information hiding principle and allows software to be designed for changeability. Every information-hiding module has an interface that provides the only means to access the services provided by the modules. The interface hides the module's implementation. Information hiding is used in object-oriented programming.
Software inspections	His approach to software inspections is quite distinct from the well-known Fagan inspection methodology. The reviewers are required to take an active part in the inspection and are provided with a list of questions by the author. The reviewers are required to provide documentation of their analysis to justify the answers to the individual questions. The inspections involve the production of mathematical tables.
Predicate logic	He introduced an approach to deal with undefined values[3] in predicate logic expressions. The approach is quite distinct from the well-known logic of partial functions developed by Cliff Jones [Jon:90].
Teaching	He has taught at various universities including McMaster University and Queens University in Canada.
Industry contributions	His industrial contribution is impressive including work on defining the requirements of the A-7 aircraft and the inspection of safety critical software for the automated shutdown of the nuclear power plant at Darlington.
Ethics for software engineers	He has argued that software engineers have a professional responsibility to build safe products, to accept individual responsibility for their design decisions, and to be honest about current software engineering capabilities. He applied these principles in arguing against the Strategic Defence Initiative (SDI) of the Reagan administration in the mid-1980s.

Table 8.1. Parnas's Achievements.

---

[3] His approach allows undefinedness to be addressed in predicate calculus while maintaining the 2-valued logic. A primitive predicate logic expression that contains an undefined term is considered false in the calculus. This is an unusual way of dealing with undefinedness and I consider his approach to be unintuitive. However, it does preserve the 2-valued logic.

8.3 Tabular Expressions

Tables of constants have used for millennia to define mathematical functions. The tables allow the data to be presented in an organized form that is easy to reference and use. The data presented in tables provide an explicit definition of a mathematical function, and the computation of the function for a particular value may be easily done. The use of tables is prevalent in schools where primary school children are taught multiplication tables and high school students refer to sine or cosine tables. The invention of electronic calculators may lead to a reduction in the use of tables as students may compute the values of functions directly from the electronic devices.

Tabular expressions are a generalization of tables in which constants may be replaced by more general mathematical expressions. Conventional mathematical expressions are a special case of tabular expressions. Conversely, everything that can be expressed as a tabular expression can be represented by a conventional expression. Tabular expressions may represent sets, relations, functions and predicates, and conventional expressions. A tabular expression may also be represented by a conventional expression but the advantage is that the tabular expression is generally easier to read and use since a complex conventional expression is replaced by a set of simpler expressions. Tabular expressions have been applied to precisely document the system requirements.

Tabular expressions are invaluable in defining a piecewise continuous function as it is relatively easy to demonstrate that the definition is consistent and that all cases have been considered. However, the conventional definition of a piecewise continuous function makes it easy to miss a case or to give an inconsistent definition. The evaluation[4] of a tabular expression is easy once the type of tabular expression is known, as each table has rules for evaluation. Tabular expressions have been applied to practical problems including the precise documentation of the system requirements of the A-7 aircraft described in [Par:01].

The discovery that tabular expressions may be employed to solve practical industrial applications led to a collection of tabular expressions that are employed to document the system requirements. Initially, little attention was given to the meaning of the tabular expressions as they worked defectively for the problem domains. However, in later work Parnas [Par:92] considered the problem of giving a precise meaning to each type of identified tabular expression in terms of their component expressions. One of Parnas's former colleagues (Janicki) at McMaster University in Canada observed that the tabular expressions identified by Parnas were members of a species, and he proposed a more general classification scheme to cover the species and to potentially identify new members in the species. He proposed a more general model of tabular expres-

---

[4] The evaluation of some of the tabular expressions is not obvious unless the reader is familiar with the rules of evaluation for the particular tabular expression.

sions [Jan:97],and this approach was based on diagrams using an artificial cell connection graph to explain the meaning of the tabular expressions. Parnas and others have proposed a general mathematical foundation for tabular expressions.

The function $f(x,y)$ is defined in the tabular expression below. The tabular expressions consists of headers and a main grid. The headers define the domain of the function and the main grid gives the definition. It is easy to see that the function is defined for all values on its domain as the headers are complete. It is also easy to see that the definition is consistent as the headers partition the domain of the function.

The evaluation of the function for a particular value (x,y) involves determining the appropriate row and column from the headers of the table and computing the grid element for that row and column.

	$y = 5$	$y > 5$	$y < 5$	H_2

H_1	$x \geq 0$	0	y^2	$-y^2$	G
	$x < 0$	x	$x+y$	$x-y$	

Fig. 8.1. Tabular Expressions (Normal Table)

For example, the evaluation of $f(2,3)$ involves the selection of row 1 of the grid (as $x = 2 \geq 0$ in H_1) and the selection of column 3 (as $y = 3 < 5$ in H_2). Hence, the value of $f(2,3)$ is given by the expression in row 1 and column 3 of the grid: i.e., $-y^2$ evaluated with $y = 3$ resulting in -9. The table simplifies the definition of the function. Tabular expressions have been employed in practical industrial projects to:

Applications of Tabular Expressions
Specify requirements
Specify module interface design
Description of implementation of module
Mathematical software inspections

Table 8.2. Applications of Tabular Expressions

Examples of Tabular Expressions

The objective of this section is to illustrate the usefulness of tabular expressions by providing a comprehensive set of examples. The examples illustrate some of the power of tabular expressions, although the examples presented here will be limited to 2-dimensional tables.

The more general definition of tabular expressions allows for multi-dimensional tables including multiple headers, and supports rectangular and nonrectangular tables. The examples presented here will usually include two headers and one grid, and the meaning of the tables is defined informally. The role of the headers and grid will become clearer in the examples.

Usually, the headers contain predicate expressions, whereas the grid usually contains terms. However, the role of the grid and the headers change depending on the type of table being considered.

Normal Function Table

The first table that we discuss is termed the normal function table and this table consists of two headers (H_1 and H_2) and one grid G. The headers are predicate expressions that partition the domain of the function; header H_1 partitions the domain of y whereas header H_2 partitions the domain of x. The grid consists of terms. The function $f(x,y)$ is defined by the following table:

$x < 0$	$x = 0$	$x > 0$	H_2

H_1		$x < 0$	$x = 0$	$x > 0$	
	$y < 0$	$x^2 - y^2$	$x^2 - y^2$	$x^2 + y^2$	
	$y = 0$	$x - y$	$x + y$	$x + y$	G
	$y > 0$	$x + y$	$x + y$	$x^2 + y^2$	

Fig. 8.2. Normal Table

The evaluation of the function $f(x,y)$ for a particular value of x,y is given by:

1. Determine the row i in header H_1 that is true.
2. Determine the column j in header H_2 that is true.
3. The evaluation of $f(x,y)$ is given by $G(i,j)$.

For example, the evaluation of $f(-2,5)$ involves row 3 of H_1 as y is 5 (> 0) and column 1 of header H_2 as x is -2 (< 0). Hence, the element in row 3 and column 1 of the grid is selected (i.e., the element $x + y$). The evaluation of $f(-2,5)$ is $-2 + 5 = 3$.

The usual definition of the function $f(x,y)$ defined piecewise is:

$$f(x,y) = x^2 - y^2 \quad \text{where } x \leq 0 \wedge y < 0;$$
$$f(x,y) = x^2 + y^2 \quad \text{where } x > 0 \wedge y < 0;$$
$$f(x,y) = x + y \quad \text{where } x \geq 0 \wedge y = 0;$$
$$f(x,y) = x - y \quad \text{where } x < 0 \wedge y = 0;$$
$$f(x,y) = x + y \quad \text{where } x \leq 0 \wedge y > 0;$$
$$f(x,y) = x^2 + y^2 \quad \text{where } x > 0 \wedge y > 0.$$

The danger with the usual definition of the piecewise function is that it is more difficult to be sure that every case has been considered as it is easy to miss a case or for the cases to overlap. Care needs to be taken with the value of the function on the boundary as it is easy to introduce inconsistencies. It is straightforward to check that the tabular expression has covered all cases and that there are no overlapping cases. This is done by examination of the headers and the headers to check for consistency and completeness. The headers for the tabular representation of $f(x,y)$ must partition the values that x and y may take and this may be done by an examination of the header.

Normal relation tables and predicate expression tables are interpreted similarly to normal function tables except that the grid entries are predicate expressions rather than terms as in the normal function table. The result of the evaluation of a predicate expression table is a Boolean value of true or false, whereas the result of the evaluation of the normal relation table is a relation. A characteristic predicate table is similar except that it is interpreted as a relation whose domain and range consist of tuples of fixed length. Each element of the tuple is a variable and the tuples are of the form $(('x_1, 'x_{2,...}, 'x_n), (x_1', x_2', ...,.x_n'))$.

Inverted Function Table

The second table that is considered is the inverted function table. This table is different from the normal table in that the grid contains predicates, and the header H_2 contains terms. The function $f(x,y)$ is defined by the following inverted table:

	$x + y$	$x - y$	xy		H_2
H_1	$y < 0$	$x < 0$	$x = 0$	$x > 0$	
	$y = 0$	$x > 0$	$x < 0$	$x = 0$	G
	$y > 0$	$x = 0$	$x < 0$	$x > 0$	

Fig. 8.3. Inverted Table

The evaluation of the function $f(x,y)$ for a particular value of x,y is given by:

1. Determine the row i in header H_1 that is true.
2. Select row i of the grid and determine the column j of row i that is true.
3. The evaluation of $f(x,y)$ is given by $H_2(j)$.

For example, the evaluation of $f(-2,5)$ involves the selection of row 3 of H_1 as y is 5 (> 0). This means that row 3 of the grid is then examined and as x is -2 (< 0) column 2 of the grid is selected. Hence, the element in column 2 of H_2 is selected as the evaluation of $f(x,y)$ (i.e., the element $x - y$). The evaluation of $f(-2,5)$ is therefore $-2 - 5 = -7$.

The usual definition of the function $f(x,y)$ defined piecewise is:

$$
\begin{aligned}
f(x,y) &= x+y & &\text{where } x < 0 \wedge y < 0; \\
f(x,y) &= x-y & &\text{where } x = 0 \wedge y < 0; \\
f(x,y) &= xy & &\text{where } x > 0 \wedge y < 0; \\
f(x,y) &= x+y & &\text{where } x > 0 \wedge y = 0;
\end{aligned}
$$

$$f(x,y) = x\text{-}y \qquad \text{where } x < 0 \wedge y = 0;$$
$$f(x,y) = xy \qquad \text{where } x = 0 \wedge y = 0;$$
$$f(x,y) = x+y \qquad \text{where } x = 0 \wedge y > 0;$$
$$f(x,y) = x\text{-}y \qquad \text{where } x < 0 \wedge y > 0;$$
$$f(x,y) = xy \qquad \text{where } x < 0 \wedge y > 0.$$

Clearly, the tabular expression provides a more concise representation of the function. The inverted table arises naturally when there are many cases to consider but only a few distinct values of the function. The function $f(x,y)$ can also be represented in an equivalent normal function table. In fact, any function that can be represented by an inverted function table may be represented in a normal function table and vice versa.

Inverted predicate expression tables and inverted relation tables are interpreted similarly to inverted function tables except that the header H_2 consists of predicate expressions rather than terms. The result of the evaluation of an inverted predicate expression table is the Boolean value true or false, whereas the evaluation of an inverted relation table is a relation.

Vector Function and Mixed Vector Table

The next table that is considered is the vector function table. This table is different from the normal table and inverted table in that the evaluation involves selecting a column in the grid. Header H_2 contains predicates and determines which column should be selected in the grid. The contents of header H_1 are variables and the result of the evaluation also involves an assignment to these variables.

	$x < 0$	$x = 0$	$x > 0$	H_2
$y_1 =$	2	0	$x+1$	
$y_2 =$	4	1	$x+2$	G
$y_3 =$	2	0	$x-1$	

H_1 is the left header containing $y_1 =$, $y_2 =$, $y_3 =$.

Fig. 8.4. Vector Table

The evaluation of the function $f(x)$ for a particular value of x is given by:

1. Determine the column j in header H_2 that is true.
2. Select column j of the grid. This yields a column (or tuple of values).
3. The evaluation of $f(x)$ is given by the tuple of values derived from the grid. The variables y_1, y_2, and y_3 are assigned to the tuple of values.

The usual definition of the function $f(x)$ defined piecewise is:

$$f(x) = (2,4,2) \qquad \text{where } x < 0;$$
$$f(x) = (0,1,0) \qquad \text{where } x = 0;$$
$$f(x) = (x+1,x+2,x-1) \qquad \text{where } x > 0.$$

The elements in the grid of a mixed vector table may be predicate expressions or terms. It is similar to the vector table in that the result of the evaluation involves the selection of a column from the grid. Header H_2 contains predicates and determines which column should be selected. The contents of header H_1 are variables and they are followed by = or |.

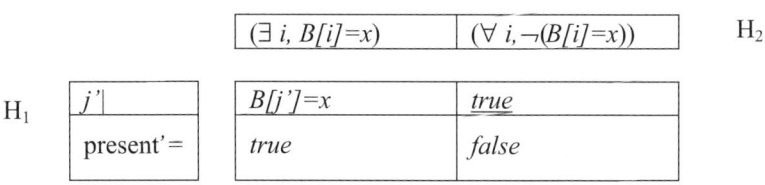

	$(\exists\, i,\, B[i]=x)$	$(\forall\, i,\, \neg(B[i]=x))$	H_2
$j'|$	$B[j']=x$	*true*	
present'=	*true*	*false*	

H_1

Fig. 8.5. Mixed Vector Table

The evaluation of the function $f(x)$ for a particular value of x is given by:

1. Determine the column j in header H_2 that is true.
2. Select column j of the grid. This yields a column (or tuple of values).
3. The evaluation of $f(x)$ is given by the tuple of values derived from the grid. The variables that are followed by = are assigned to the corresponding values in the column of values. The variables followed by | yield a relation such that the predicate expression is true in the column.

The example above is concerned whether the element x appears in the array B. Suppose x occurs in position 3 and 5 of array B. Then the result is that present is set to true and the relation $(j,3)(j,5)$ indicates the places in the array B where the value x appears.[5]

Generalized Decision Table

The next table that is considered is the generalized decision table. This table is different from the tables considered so far in that the evaluation involves first substituting the entries in header H_2 into the main grid. The result of the substitution is that one row is true in the grid.

---

[5] I considers the normal table and inverted table to be reasonably intuitive but am of the view that the mixed vector table as unintuitive.

$x + y$	xy	$x + 2y$	
			H_2

		$1 < \#$	$1 = \#$	$1 > \#$	
H_1	xy				G
	$x + 3y$	$2 < \#$	$2 = \#$	$2 > \#$	
	x^2	$3 < \#$	$3 = \#$	$6 > \#$	

Fig. 8.6. Generalized Decision Table

The evaluation of the function $f(x,y)$ for a particular value of x,y is given by:

1. Substitute the terms in each column of header H_2 for the # symbol in the column of the grid.
2. Determine the row i of the grid that is true.
3. The evaluation of $f(x,y)$ is given by $H_1(i)$.

The evaluation involves $x + y$ being substituted for the # symbol in column 1 of the grid; xy being substituted for the # symbol in column 2 of the grid; and $x + 2y$ being substituted for the # symbol in column 3 of the grid. Consider the evaluation of $f(x,y)$ for $x = 3$, $y = 1$. Then $x + y$ is 4; xy is 3; and $x + 2y$ is 5. Thus row 3 is the only row of the grid in which all of the predicate expressions evaluate to true. Thus $f(3,1)$ is given by the evaluation of $H_1(3) = 3^2 = 9$.

NRL Mode Transition Tables

The next table considered is the mode transition table and this table arose in practice from an examination of the transitions of a system mode to a new mode following the occurrence of a particular event.

H_1	G	H_2
Current Mode	**Event**	**New Mode**
Too Low	T_E(WaterPres $\geq$ Low)	Permitted
Permitted	T_E(WaterPres $\geq$ Permit)	High
Permitted	T_E(WaterPres $<$ Low)	Too Low
High	T_E(WaterPres $<$ Permit)	Permitted

Fig. 8.7. Mode Transition Table

The notation T_E(WaterPres $\geq$ Low) means that the event of water pressure being equal to or exceeding the value of "Low" has occurred. The evaluation of the function for a particular mode m is given by:

1. Identify the rows (i, j, k …) that current mode m occurs in H_1.
2. Identify the row r in (i, j, k …) in G that corresponds to the event that has occurred.
3. The new mode is given by $H_2(r)$.

Suppose the current mode is Permitted and the event T_E(WaterPres < Low) occurs. Then rows 2 and rows 3 of header H_1 correspond to the current mode and the event corresponds to row 3. Thus the new mode is 'Too Low' as given by $H_2(3)$.

An equivalent formulation of the above mode transition table is given by a similar NRL mode transition table.

H_3

Current Mode	(WaterPres ≥ Low)	(WaterPres ≥ Permit)	(WaterPres < Low)	(WaterPres < Permit)	New Mode
Too Low	T_E	-	-	-	Permitted
Permitted	-	T_E	-	-	High
Permitted	-	-	T_E	-	Too Low
High	-	-	-	T_E	Permitted

H_1 G H_2

Fig. 8.8. NRL Mode Transition Table

The evaluation of the function for a particular mode m is given by:

1. Identify the rows $(i, j, k ...)$ that current mode m occurs in header H_1. (Usually, only one row will apply).
2. Each row in the grid consists of one or more (event) functions of one argument (as in H_3). The rows $(i, j, k ...)$ are then applied to the values in H_3 (the argument of the function g_p in column p is determined from the value in column p of H_3).
3. This result are rows $(i, j, k ...)$ where each row is a one dimensional grid of Boolean values.
4. The logical or operation is then applied to the Boolean values in each row to yield a Boolean value for each row.
5. This results in one row r that is true and the new mode is given by $H_2(r)$.

An example should help to make the above clearer. Suppose the current mode is Permitted as before and that the event T_E(WaterPres < Low) occurs. Then rows 2 and 3 of H_1 are applicable as the current mode is Permitted, and rows 2 and 3 of the grid are selected.

Next, rows 2 and 3 of the grid are applied to the values in H_3 to yield 2 rows of 1-dimensional grids of Boolean values as below.[6]

F	T_E(WaterPres $\geq$ Permit)	F	F
F	F	T_E(WaterPres < Low)	F

Fig 8.9. Result of Function Application of Grid to H_3

Next, the logical 'or' operation is then applied to the Boolean values to yield a Boolean value for each row and this results in:

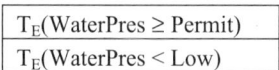

Fig 8.10. Result of Logical 'or' of rows

These are the results from row 2 and row 3, and as the event T_E(WaterPres < Low) has occurred row 3 is the row to apply and the new mode is then given by H_2 (3), i.e., Too Low.

8.4 Software Development Documentation

Most mature software companies view documentation as an essential part of the software development process. Companies may develop their own documentation standards or use existing standards (e.g., IEEE 1074 for software development and IEEE 1059 for software verification and validation). The objective of the documentation of the requirements is to ensure that all parties (the customer, developers, and testers) share a common understanding of the proposed system, and are therefore in a position to implement the system and verify its correctness. The documentation of the design demonstrates how the system will be implemented to satisfy the requirements. The developers will implement the system in a programming language and software inspections verify the correctness of the implementation with respect to the design and requirements. The objective of the test documentation is to specify the test cases to be used in testing to verify that the software satisfies the requirements.

The documentation employed by most companies (including the IEEE standards) is informal and offers no mechanism to rigorously demonstrate that the design satisfies the requirements or that the code satisfies the design. Instead, companies employ inspections [Fag:76, Glb:94] carried out by experienced personnel and the objective is to verify that the documentation is correct and satisfies the requirements, and to identify defects in the requirements at the early stages of the lifecycle.

Other disciplines such as engineering regard mathematics as an essential part of their work and most engineering documents include a substantial

---

[6] I consider the evaluation of the NRL table to be cumbersome and view the table as unintuitive.

amount of mathematics. This has led to interest in an engineering approach to software documentation and to the use of mathematics for software documentation. The use of mathematics allows greater precision in the expression of the requirements and allow rigorous software inspection [Par:94]. The following mathematical documents have been proposed by Parnas in [Par:95]:

Mathematical Documents
Systems requirements document
System design
Software requirements
Module interface specification
Module internal design

Table 8.3. Mathematical Documents

Each document has a set of objectives and addresses the needs of a specific audience. The objective of the requirements document is to ensure that all parties have a precise mathematical understanding of the requirements of the system. The mathematics employed allows a more rigorous demonstration of correctness of the design than that achieved by natural language. Good documentation is used as a reference throughout the software development.

Most mature software companies expect software documents to be kept up to date and to reflect the actual software product. However, in practice [Let:03], the documentation is often out of date and this makes maintenance more difficult. Mathematical documentation has been used in a number of software projects, but its use is very much the exception rather than the general rule. Some reasons for this include:

No.	Low level of use of Mathematical Documentation
1.	Many software engineers receive an elementary education in mathematics.
2.	Many customers have a limited background in mathematics and have a limited interest in learning mathematics.[7]
3.	The conventional wisdom has been to use a natural language such as English for writing documents.
4.	A proven usable and cost effective approach to mathematical documentation (that has minimal impact on project cost and schedule) has not been widely communicated.
5.	There are no standards for mathematical software documentation.

Table 8.4. Reasons for Low Level of User of Mathematical Documents

---

[7] I recall mentioning that this was my experience to Parnas only to find that Parnas had different experiences. However, I see my industrial experience as a little more comprehensive than Parnas's, and the exceptions that Parnas is familiar with really prove the rule.

This section describes the Parnas approach to mathematical documentation [Par:95]. The only way that mathematical documentation will achieve general use for future software projects is if it can be demonstrated to be cost effective and that superior results will be achieved by their use.[8] It needs to be shown that the deployment of mathematical documentation leads to higher quality software with no significant adverse effect on the project schedule or cost, as these are key drivers in most companies. Comprehensive empirical studies are required to evaluate whether mathematical software documentation achieves these goals.

The steps required to make mathematical documentation a reality in a company include:

No.	Step
1.	Define a suite of standards for mathematical software documentation.
2.	Define the process for using mathematical documentation and develop a case study with samples of the completed mathematical documents.
3.	Provide practical training on the mathematical standards and on the process for mathematical documentation.
4.	Conduct a pilot(s) of the standards and process within the company and gather data on the effectiveness of the mathematical approach. The results of the pilot need to be communicated within the company. There may be a need for some revisions to the standards or process.
5.	Provide support and training to the project team should difficulties arise with implementation of the standards in early projects. This could potentially involve reviews by an outside consultant to ensure that the mathematical documents are correct and achieve their purpose.

Table 8.5. Steps to Introduce Mathematical Documentation

Overview of Mathematical Documentation

The following mathematical documents are proposed by Parnas [Par:95] to be part of the software development process:

---

[8] Superior results means higher quality and software reliability, faster time to market, reduced costs of development, and increased productivity. These need to be measured carefully before a judgment is made as to whether a particular method has real added value. There have been some encouraging results from the use of tabular expressions as a method for the precise documentation of software. However, industrials will require clear quantitative data to make judgments on whether to pilot the methodology. The bottom line is that the mathematical approach should be a more cost effective solution than conventional approaches. I recall asking Parnas for empirical data to demonstrate that his method achieved the superior results required.

Document	Description
System requirements	This is a black box description of the system and identifies the environment and quantities of concern to the users and the constraints to be enforced by the system. It associates a mathematical variable with each quantity of concern to the user.
System design	This document describes the computers in the system and describes how they communicate. It defines the relationship between the values in the input and output register of each computer and the environmental quantities identified in the system requirements document.
Software requirements	The software requirements document is determined from the system requirements document and the system design document. It specifies the input-output behaviour of the software.
Module guide	This document describes the division of software into modules, and describes the responsibilities of each module. The software modules must satisfy the requirements.
Module interface specification	The module interface specification provides a black box description of each module listed in the module guide. It describes the externally-visible effects of using the module, and does not include implementation details.
Internal design	This document describes the data structure of each module and specifies the effect of each access program on the data structure. The module internal design is a refinement of the module interface specification.
Uses-relation document	This consists of a relation where the pair (P,Q) is in the relation if program P of a module uses program Q of a module. The range and domain of the uses relation are subsets of the set of access-programs of the modules.
Software test specification	This document describes the tests that will be conducted in order to verify the correctness of the software.
Other documents	Other documents that may potentially be employed as part of the software process include data-flow documents, service specification document, protocol design document, and the chip behaviour specification document.

Table 8.6. Mathematical Documentation

Some of the key principles in the Parnas approach include:

Key Principles of Mathematical Documents
Top down design and black box specification.
Modularity is in accordance with the information-hiding principle.
Verification of the design decisions against black box specification.
Use of classical mathematics and predicate logic.
Use of tabular expressions to represent functions.
The mathematical documents are referred to during software development.
The documents are placed under strict configuration management control.

Table 8.7. Key Principles for Mathematical Documents

8.5 System Requirements

The Parnas approach to the specification of the system requirements is to employ mathematical relations and tabular relations. An early approach was based on a 2-variable model and this involved generating a list of all the outputs from the computer and another list of all the inputs to the computer. The next step involves writing descriptions of mathematical functions that mapped the (history of) values of the input variables to values of the output variables.

Environment

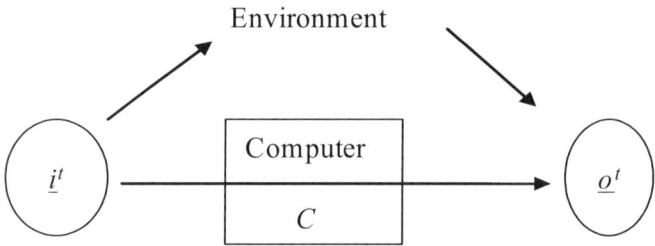

$$\underline{o}^t = C(\underline{i}^t) \text{ for some computer } C$$

Fig. 8.11. Two Variable Model

The simple model proved to be too simple in practice as the mappings from the inputs (values obtained from special purpose hardware) to outputs (values that controlled specialized hardware) proved to be very complex in some cases. The inputs and outputs from the computer were not visible to, or even of interest to, the users. This led to a reexamination and revision of the model. [9]

---

[9] A black box specification for the system requirements is slightly cumbersome and it involves defining responses in terms of the stimuli history. Often, the output is dependent on state data stored in the computer as well as the inputs themselves, and I prefer the state-based approach to specifica-

The 2-variable model evolved over time to become a 4-variable model. In addition to the inputs and outputs produced by the software, there are variables representing the real interests of the users. These variables are termed *monitored and controlled environmental variables*. These are the inputs and outputs that the user of the system is aware of and interested in.

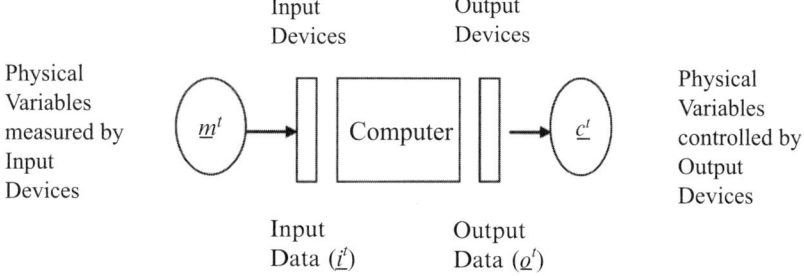

Fig. 8.12. The 4-Variable Model

The 4-variable model includes the use of four types of variables: monitored, controlled, input, and output; and several relations including: REQ which expresses the requirements of the system; NAT which expresses the constraints on the values of the environment variables due to restrictions imposed by nature or previously installed systems; the relation IN that specifies the input to the computer; the relation SOF that specifies the behavior of the software, and the relation OUT that specifies the output from the computer. The precise mathematical formulation is described below:

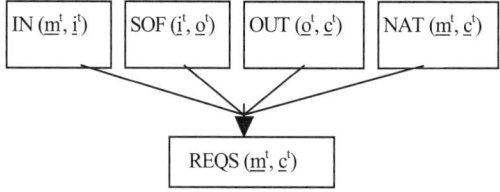

Fig. 8.13. Acceptable Software Behavior

The diagram expresses the Parnas formula for the acceptability of the software. The precise mathematical formulation is given by:

$$\forall \underline{m}^t \ \forall \underline{i}^t \ \forall \underline{o}^t \forall \underline{c}^t$$

---

tion as employed in *Z* or VDM. The response to a stimulus in the state-based approach involves a change of state.

$$[\text{IN } (\underline{m}^t, \underline{i}^t) \wedge \text{SOF } (\underline{i}^t, \underline{o}^t) \wedge \text{OUT } (\underline{o}^t, \underline{c}^t) \wedge \text{NAT } (\underline{m}^t, \underline{c}^t) \Rightarrow \text{REQS } (\underline{m}^t, \underline{c}^t)]^{10}$$

This formula seems very complicated and seems to bear little relationship to requirements specification. However, the formula is essentially composed of five mathematical relations IN, SOF, OUT, NAT, and REQS and in essence describes the behavior that the software must exhibit to be acceptable for use and for the system requirements to be satisfied. The terminology and individual relations are described in detail later in this chapter.

In practice, a formal proof of this equation is rarely needed (or practical). Instead, it may be used to generate a set of checklists or testing criteria. The mathematical documentation of the system requirements is discussed below.

Monitored Variables

These are the externally visible variables that the system needs to measure. They are variables that the user of the system is aware of and include input data typed by a user, the physical inventory monitored by an inventory system, account balances, or the state of a switch. These environment variables are categorized into those that are specific to the application, those that are common to hardware devices, and those that are specific to hardware devices.

The collection of monitored variables is denoted by $m_1, m_2, \ldots, m_n$. Each of these monitored variables can be expressed as a function of time m_i^t and the vector of time functions of monitored variables $(m_1^t, m_2^t, \ldots, m_n^t)$ is denoted by $\underline{m}^t$. A description of the physical quantity (including the physical interpretation) is given for each monitored environment variable. This includes the name, set of possible values, and data type of each monitored environment variable.

Mon Variable	Description	Type
EntBadgeId	Determines the numeric badge id from the electronic encoding of the badge.	Numeric
ValidBadgeIds	Records the list of valid badges.	Set of numeric

Table 8.8. Monitored Environment Variables

Controlled Variables

These are quantities that the user of the computer system wishes to compute and control such as signals to control external hardware (e.g., the setting of a switch that controls the opening or closing of a door to "on" or "off" depending on whether the owner of the badge has sufficient authorization privileges to enter the room). They also include variables that control the displayed balance of an account (e.g., a credit card account) on a screen. They are usually continuous

---

[10] The condition for the acceptability of the software is an elegant but complex formula and it is debatable as to how easy it is to work with in practice. Parnas has extended the 4-variable model to a 5-variable model. However, I wonder how all this complexity helps.

real valued variables that exist outside the system and they control the computation of values, the settings of external devices, and the display of data on reports or computer screens. The values of the controlled variables are derived from operations involving the values of the monitored variables. The environment variables are categorized into those that are specific to the application, those that are common to hardware devices, and those that are specific to hardware devices.

The collection of controlled variables is denoted by $c_1, c_2,...,c_k$. Each of these controlled variables can be expressed as a function of time c_i^t and the vector of time functions of controlled variables $(c_1^t, c_2^t,..., c_k^t)$ is denoted by $\underline{c}^t$. A description of the physical quantity (including a physical interpretation) is given for each controlled variable. It includes the data type and the set of all possible values of each controlled variable.

The definition of DoorSetting is below. Its value determines whether the door will be opened in response to the card being swiped. The second example is a controlled variable that displays the savings account balance on the screen.

Ctr Variables	Description	Type
DoorSetting	The value of DoorSetting determines whether the door will be opened.	Open or Closed
DispSavAccBal	Displays the savings account balance on the screen	Numeric

Table 8.9. Controlled Environment Variables

Environment Restrictions (NAT)

The environment may place constraints on the values of the controlled and monitored variables. These constraints are expressed by defining the externally visible input and output histories that can occur. These are recorded in the relation NAT (i.e., Nature). They express the restrictions imposed by nature (or by previously installed systems) on the values of the environmental quantities. The system requirements for the proposed system (i.e., REQ defined below) will need to be feasible with respect to these constraints.

1. Dom(NAT) is a set of vector valued time functions containing the possible values of $\underline{m}^t$.
2. Ran(NAT) is a set of vector valued time functions containing the possible values of $\underline{c}^t$.
3. $(\underline{m}^t, \underline{c}^t) \in$ NAT if the constraints allow the controlled variables to take on the values described by $\underline{c}^t$ when the values of the monitored variables are described by $\underline{m}^t$.

Behavioral Requirements (REQ)

The behavioral requirements of the computer system are given by the binary relation REQ. This expresses the externally visible input and output histories that the system should permit. There is usually one relation for each element of C and occasionally one relation for all of C. The domain is a set of values for M and the range is a set of values for C. Tabular notation is employed to define the value of each controlled variable as a function of monitored variables. The relation REQ is defined mathematically as:

1. Dom(REQ) is a set of vector valued time functions containing the permissible values of $\underline{m}^t$
2. Ran(REQ) is a set of vector valued time functions containing the permissible values of $\underline{c}^t$.
3. The pair $(\underline{m}^t, \underline{c}^t) \in$ REQ if the computer system allows the controlled variables to take on the values described by $\underline{c}^t$ when the values of the monitored variables are described by $\underline{m}^t$.

The feasibility of the requirements REQ with respect to the environment constraints NAT needs to be verified. This requires that behavior is specified for any input sequence that could occur. That is, for any given situation that NAT says can arise, then REQ specifies an action that is allowed by NAT.

1. Check dom(REQ) $\supseteq$ dom(NAT)
2. Check dom (REQ $\cap$ NAT) = dom(REQ) $\cap$ dom(NAT)[11].

Mode Definitions

A mode class corresponds to a partitioning of the set of system states. There may be several mode classes associated with the system. There are several modes in each mode class, and each mode corresponds to a set of system states. At any given moment of time the system is in exactly one mode from each mode class, and initially the system is in the initial mode of each mode class. Events trigger changes to the system mode, and the new mode is given by the mode transition function.

The example below is concerned with water pressure in a water plant system or a hydroelectric scheme. The water pressure within the plant may be too low, within the permitted range or too high. The system has one mode class (M_c_Operating) and there are three modes in the mode class. These modes reflect that the water level is too low, at the permitted level, or too high.

---

[11] Condition 2 can be expressed as dom (REQ $\cap$ NAT) = dom(NAT) since dom(REQ) $\supseteq$ dom(NAT).

Mode Class	Modes in Class	Initial Mode
M_c_Operating	M_d_Too Low M_d_Permitted M_d_High	M_d_Permitted

Table 8.10. Mode Classes and Modes

Events will cause a change in mode to occur; for example, a sudden drop in water pressure will cause a mode transition from permitted to too low. The mode transitions is defined by a mode transition table. The notation $E_{(A)}$ indicates that the event A has occurred.

Old Mode	Event	New Mode
M_d_Too Low	$E_{(M\ WaterPres \geq Low)}$	M_d_Permitted
M_d_Permitted	$E_{(M\ WaterPres \geq Permit)}$	M_d_High
M_d_High	$E_{(M\ WaterPres < Permit)}$	M_d_Permitted

Table 8.11. Mode Transition Table

Exceptions and Undesired Events

Exceptions and undesired events that may occur need to be considered as otherwise an improper response may result if designers have not considered what should happen in these cases.

Timing and Accuracy Constraints

Precision and tolerance of the controlled and monitored variables needs to be considered as these are continuous real valued numbers that are subject to the variations in measurement that arise with physical devices. The ideal behavior is specified in NAT and REQ, and the accuracy constraints specify the permitted deviation from the ideal behavior.

The precision function is described on the individual monitored variables and a tolerance function is specified on the individual controlled variables. A precision function on the vector of monitored variables $\underline{m} = (m_1, m_2,..., m_n)$ is given by $\underline{P}$ where $\underline{P} = (P_1, P_2,..., P_n)$ and P_i is the precision function of the monitored variable m_i. Similarly, the tolerance function can be specified for the control vector $\underline{c} = (c_1, c_2,..., c_m)$ and is given by $\underline{T}$ where $\underline{T} = (T_1, T_2,..., T_n)$ and T_i is the tolerance function of controlled variable c_i.

Timing constraints are expressed using classical mathematics.[12] Let t be a time, and EC an event class. $EC(P) = \{(s,t) : P(s,t)\}$ where P is a predicate and $s(t)$ is the environment state function. The time of the next event to occur is :

---

[12] Parnas prefers to avoid using the temporal operators introduced by others and classical mathematics enables the timing to be dealt with effectively.

$$\text{Next}(EC,t) = t', \; t' \text{ is the smallest time such that } (t' > t) \wedge (s,t) \in EC$$
$$= \infty \; (\text{if no such } t' \text{ exists }).$$

Assumptions and Expected Changes

The fundamental assumptions and likely changes to the system need to be identified early.

8.6 System Design and Software Requirements

The system design phase is concerned with the design decisions that determine the number of computers, the physical hardware, and the nature of the interconnections between these computers. There are two additional sets of variables: a set of inputs (i.e., variables that can be read by the computers in the system) and a set of outputs (i.e., variables whose values are determined by the computers in the system).

Step	Description
1.	Identify the inputs to the computer (i.e., the input alphabet).
2.	Identify the outputs of the computer (i.e., the output alphabet).
3.	Define the relation IN • Dom(IN) is a set of vector valued time functions containing the possible values of $\underline{m}^t$ • Ran(NAT) is a set of vector valued time functions containing the possible values of $\underline{i}^t$ • $(\underline{m}^t, \underline{i}^t) \in$ IN if $\underline{i}^t$ describes possible values of the input variables when $\underline{m}^t$ describes the values of the monitored variables.
4.	Define the relation OUT. • Dom(OUT) is a set of vector valued time functions containing the permissible values of $\underline{o}^t$ • Ran(OUT) is a set of vector valued time functions containing the permissible values of $\underline{c}^t$ • $(\underline{o}^t, \underline{c}^t) \in$ OUT if the computer system allows the controlled variables to take on the values described by $\underline{c}^t$ when $\underline{o}^t$ describes the values of the output variables.
5.	Check the validity of IN with respect to NAT: (i.e., dom (IN) $\supseteq$ dom(NAT)).

Table 8.12. System Requirements

The software requirements are determined from the system requirements and the systems design. The software requirements (input-output behavior of the software) are defined in the relation SOF.

Step	Description
1.	Dom(SOF) is a set of vector valued time functions containing the possible values of i^t
2.	Ran(SOF) is a set of vector valued time functions containing the possible values of o^t
3.	$(i^t, o^t) \in$ SOF iff the software can take on the values described by o^t when i^t describes the values of the input variables.

Table 8.13. Software Requirements

The relation SOF is redundant as it may be derived from IN, OUT and REQ.

$$SOF = OUT^{-1} \circ (REQ \circ IN^{-1}).$$

8.7 Software Design

Many software projects are too large (e.g., >100 person years) to be completed by a single person in a short period. Therefore, the software needs to be divided into a number of modules (components) that can be developed separately in order to produce the software product in a timely manner. Each module is a work product for a person or team. The actual decomposition is usually decided by management and involves assigning staff to carry out project work based on the expertise available.

Large programs are generally more complex than smaller programs with several programmers involved, lots of tiny details to be remembered, and no one person knowing everything about the program. Communication becomes more difficult in larger teams, and care is needed to ensure that if a change to the requirements or design occurs that all affected parties are informed and aware of the change.

The division of a product into separate parts is an essential part of manufacturing, and the component parts are then combined to form the product. The components may be manufactured in separate locations. Hardware components have a well-defined interface that specifies its behavior and it is clear how and when the hardware components are to be combined to form the larger product.

A software system consists of modules that are combined to form a larger system. There are several different ways to put software modules together to form the software product and they may be combined at different times (e.g., during compile time, linking of object programs or running a program in limited memory). The way that software programs may be combined places constraints on changing each module independently of the others, the names employed in the modules, and the size of memory.

Parnas revolutionized the approach to software design and module decomposition by advocating stated technical criteria to perform effective module decomposition [Par:72]. He advocates that the specification of the module should focus on what the module should do rather than how it should be done (i.e., the specification of the software module should abstract away from the implementation details). The abstract specification allows the developer the maximum amount of freedom in choosing the best implementation of the model.

The heart of the Parnas design philosophy is information hiding[13] and this allows a module to access a second module without needing to know the implementation details of the second module. This is achieved by the concept of the "secret" of the module and the module interface specification. The Parnas approach allows modules to be designed and changed independently. They may be divided into smaller modules and reused. The approach is to identify the design decisions[14] that are likely to change, and to then design a module to hide each such design decision. The approach of designing for change in requirements or system design requires knowing what these changes might be. Change therefore needs to be considered as early as possible in the requirements phase.

Information hiding is the vehicle in designing for change. Every information hiding module should have an interface that provides the only means to access the services provided by the module. The secret is a design decision that may be changed without affecting any other module. A secret may be a data structure or an algorithm. A data structure that is the secret of the module is internal to the module rather than external, and a change to the definition of the data structure changes just one module (i.e., the module of which it is the secret) and other modules are not affected. Similarly, a secret algorithm is not visible outside the module, and should the particular algorithm be changed, then there is no impact on other modules. A shared secret is not a secret and a module is not aware of the secret of any other module. Information hiding is an essential part of good class design in object-oriented programming.

Often, in poorly designed systems there are unnecessary connections. For example, code may be duplicated in the system or modules may contain code that is dependent on data structure design decisions in other modules. Clearly, these modules are not designed for changeability, as a change in one module requires changes in several other modules. The Parnas design procedure for large programs is as follows:

---

[13] I see information hiding as Parnas's greatest achievement.

[14] Of course, identifying and predicting all likely changes to the software is impossible in some domains. It would require sound product and marketing knowledge; prediction of future consumer behavior; prediction of future technology trends, etc. However, it is important to identify as many likely changes as is feasible as this will enable the software to be designed as far as is practical for change.

Step	Description
1.	Decompose the system into a set of modules with each module designed in accordance with the information hiding principle. Identify the secrets (i.e., the design decisions that are likely to change). It is essential that the secret should not be shared.
2.	Design an information-hiding interface for each secret. • Implementation of the access programs is based upon the secrets. (The access programs are externally visible to the other modules). • The interface does not change if the secret changes.
3.	Specify all of the interfaces precisely. This is done using trace specifications.
4.	Implement each module independently using only interface information. • One module may use the access programs of another module. • Program from one module cannot use the hidden data structures or internal programs from another module. •Modules should be kept small with just one secret.

Table 8.14. Designing for Change

Engineers are required to design and build products that are safe for the public to use, and a good design is essential to the delivery of a high-quality product that will meet customer expectations. The design of the product must satisfy the system requirements and the components of the product are detailed in the design. The software field has been ineffective in developing components that may be reused[15] in various applications.

The quality of the design will determine how easy it is to change the software over time. Ease of change is essential for good maintainability of the software, as often customers will identify further desirable to enable them to achieve their business objectives more effectively. Change therefore needs to be considered as early as possible.

Information hiding allows modules to be designed and changed independently. They may be divided into smaller modules and reused. Designing for change requires a focus on identifying what the likely changes might be early in the software development process. An evaluation of what changes the module decomposition can handle can be determined by comparing the list of changes the design can easily accommodate with the list of expected changes.

---

[15] Parnas has joked that we have developed all of this reusable software that is never reused.

Module Interface Design

The interface between two modules consists of all the assumptions that the developer of each module has made about the other. If the correctness of one module can only be demonstrated by making an assumption about another module, and a change in the other module makes that assumption false, then the first module will have to change. It is therefore essential to be aware of the assumptions implicit in an interface. If many parts of the software are based on an assumption in a particular interface, and the assumption is invalid, then there is a need to change many parts of the software.

The Parnas approach to interface design is to use an abstract interface rather than a concrete interface. This is an interface that models some (but not all) properties of an actual interface. It is a precise formally specified interface that is a model of all expected actual interfaces. All things that are true of the abstract interface must be true of the actual interface. Abstract interfaces involve eliminating details, especially the details that are likely to change. Abstractions are simplifications of the reality and they apply in many situations (each situation shares the abstraction but differs in things that are abstracted from).[16]

If the abstract interface is designed correctly then real world changes that affect the actual interface affect only the interface programs provided that the assumptions in the abstract interface remain valid. The abstract interface is specified using trace assertions. The design of an abstract interface involves:

Step	Description
1.	The module interface specification is a black box description of the modules. It specifies how the access programs may be accessed from other modules. The interface specification will not have to change if the program (or secret) changes.
2.	Prepare a list of assumptions about properties of all the possible real world interfaces to be encountered. Review and revise the list of assumptions.
3.	Express these assumptions as a set of access programs (operators, method). This will be the basis of a module specification.
4.	Perform consistency checks • Verify that any property of the access program set is implied by the assumptions. • Verify that all assumptions are reflected in the interface specification. • Verify that the bulk of the system can be written using these access programs.

Table 8.15. Module Interface Specification

---

[16] This is similar to the refinement of an abstract specification into a more concrete specification. There are many concrete specifications that satisfy the abstract specification, reflecting the fact that the designer has a choice in the design of the system. A valid refinement is required to satisfy the well-known commuting diagram property.

Trace Assertion Method

The trace assertions method is employed to specify the module interface specification. Each module consists of a hidden data structure (or object) and a set of access programs that may be invoked externally by other modules. The access programs are the only programs that may alter the data structure. The trace assertion method provides a black box description of the module without revealing the implementation of the module. This gives the programmer freedom in choice of implementation to satisfy the requirements of the module.

A trace consists of a sequence of events. Each event consists of the access program name and the values of its arguments. Two traces are equivalent if the resulting behavior of the module is identical for each trace. There will generally be many traces equivalent to a particular trace, and the set of traces equivalent to a particular trace is termed the equivalence class of the trace. A canonical trace is chosen as a representative of the equivalence class. A trace describes all data passed to and returned from the module, and is a way to describe effects where the visibility of the effect is delayed.

The behavior of a module interface specification is described by traces (i.e., sequences of discrete events). The trace assertion specifications comprise three groups of relations:

Relation	Description
1.	Functions whose domain is a set of pairs (canonical traces, event) and whose range is the set of canonical traces.[17] The pair $((T_1,e),T_2)$ is in the function f_E iff the canonical trace T_2 is equivalent to the canonical trace T_1 extended by the event e.
2.	Relations R_O whose domains contain all the canonical traces and whose range is a set of values of output variables.
3.	Functions f_V whose domain is the set of values of the output variables and whose range defines the information returned by the module to the user of the module.

Table 8.16. Relations in Trace Assertion Method (TAM)

Internal Design

The documentation of the module design involves producing the module guide, the module interface specification, the uses relation and the module internal design. Large systems are decomposed into modules with each module then implemented by one or more programmers. The modules are designed according to the information hiding principle.

The module guide is an informal document that describes the division of the software into modules and includes a precise and complete description of

---

[17] A canonical trace is a finite subset of the infinite set of possible traces.

each module. It makes it easy for maintainers of the system to find the module that they are looking for and to make changes or corrections. The secret of each module is detailed in the module guide. The module guide is produced by the designers and used by the maintainers. The responsibilities of each module are documented.

The module interface specification is a black box description of the modules. It specifies how the access programs may be accessed from other modules. The module internal design describes the data structure of each module and specifies the effect of each access program on the data structure. The uses relation describes the relation between the access programs of modules with pair (P,Q) in the uses relation if access program P uses access program Q.

The internal design of a module includes three types of information:

Part	Description of Internal Design
1.	A complete description of the data structure.
2.	An abstraction function f such that $(((o,d), T) \in f$ if and only if a trace equivalent to T describes a sequence of events affecting the object o that could have resulted in data state d.
3.	An LD relation[18] (known as the program function) that specifies the behavior of each program in the module in terms of mapping from data states before execution to data states after execution.

Table 8.17. Internal Design Description

8.8 Software Inspections

The Parnas approach to software inspection offers a rigorous mathematical approach to finding defects and to verifying the correctness of the software. It is quite distinct from the well-known Fagan Inspections developed by IBM in the mid-1970s. It has been successfully employed in the safety critical field (e.g., in the inspection of the shutdown software for the nuclear power plant at Darlington, Canada). The use of mathematical documents is at the heart of the procedure as the mathematics allows complete coverage.

Tabular expressions are employed to provide a systematic examination of all cases to take place. Some of the key features of Parnas inspections include [PaW:01]:

---

[18] The effect of executing a program is described by an LD relation. An LD relation consists of an ordered pair (R_L, C_L) where R_L is a relation on the set of states U and C_L is a subset of Dom R_L. C_L is termed the competence set of the LD relation. Termination is guaranteed if execution commences in a state in the competence set of the relation.

No.	Feature
1.	All reviewers are actively involved. The author (designers) pose questions to the reviewers rather than vice versa.
2.	Several types of review are identified. Each review type focuses on finding different types of errors.
3.	Each review requires specific expertise to be available.
4.	Reviewers with the right expertise are chosen. The effort of a reviewer is focused
5.	Reviewers are provided with a questionnaire and are required to study the document(s) carefully in order to find answers to the questions. They are required to justify their answers with (mathematical) documentation.
6.	The reviewers proceed systematically so that no case or section of program is overlooked.
7.	Issues raised by reviewers are discussed in small meeting with the author/designer.
8.	Most reviews are tightly focused. There is one overall review to decrease likelihood that problems have been overlooked.

Table 8.18. Characteristics of Parnas Inspections

Conventional reviews may be criticized on the grounds that they lack the rigour associated with mathematical reasoning, and do not provide a rigorous mechanism to ensure that all cases have been considered, or that a particular case is correct. The Fagan {Fag:76} or Gilb [Glb:94] style review focuses on the management aspects of software inspections including the role of the participants, the length of the meeting, and the forms for reporting. Their approach is to paraphrase a document or code in a natural language, and although these approaches yield good results, the Parnas approach offers the extra rigor that is achieved by the use of mathematics and tabular expressions. This is especially important for the safety critical field.[19]

Software inspections play a key role in building quality into a software product, and testing plays a key role in verifying that the software is correct and corresponds to the requirements. There is clear evidence that the cost of correction of a defect increases the later in the development cycle in which the defect is detected. Consequently, there is an economic argument to employing software inspections as there are cost savings in investing in quality up front rather than adding quality later in the cycle.

---

[19] Parnas inspections are unlikely to be cost effective in mainstream software engineering. I have seen no empirical data on the amount of time needed for a Parnas inspection. The results of the safety inspection of the Darlington software is impressive, but it was very time consuming (and expensive). Some managers in Darlington are now less enthusiastic in shifting from hardware to software controllers [Ger:94].

The well-known Fagan inspection methodology and the Gilb methodology include preinspection activity, an inspection meeting, and postinspection activity. Several inspection roles are employed, including an author role, an inspector role, a tester role, and a moderator role.

The formality of the inspection methodology used depends on the type of business of the organization. For example, telecommunications companies tend to employ a very formal inspection process, as it is possible for a one line of code change to create a major telecommunications outage. Consequently, a telecommunications company needs to ensure that the quality of the delivered software is fit for use, and a key part of building in the desired quality in is the use of software inspections.

The quality of the delivered software product is only as good as the quality at the end of each particular phase. Consequently, it is desirable to exit the phase only when quality has been assured in the particular phase. Software inspections assist in ensuring that quality has been built into each phase, and thus ensuring that the quality of the delivered product is good.

Fagan Approach to Inspections

The Fagan methodology is a seven-step process, including planning, overview, preparation, inspection, process improvement, rework, and follow-up activity. Its objectives are to identify and remove errors in the work products, and also to identify any systemic defects in the processes used to create the work products. A defective process may lead to downstream defects in the work products.

The process stipulates that requirement documents, design documents, source code and test plans all be formally inspected by experts independent of the author, and the experts inspect the deliverable from different viewpoints, for example, requirements, design, test, customer support, etc.

There are various roles defined in the inspection process, including the moderator, who chairs the inspection, the reader, who paraphrasing the particular deliverable and gives an independent viewpoint, the author, who is the creator of the deliverable; and the tester, who is concerned with the testing viewpoint. The inspection process will consider whether a design is correct with respect to the requirements, and whether the source code is correct with respect to the design. The Fagan inspection process is summarized by:

Step	Description
Planning	This includes identifying the inspectors and their roles, providing copies of the inspection material to the participants, and booking rooms for the inspection.
Overview	The author provides an overview of the deliverable to the inspectors.
Prepare	All inspectors prepare for the inspection and the role that they will perform.
Inspection	The actual inspection takes place and the emphasis is on finding major errors and not solutions.

Process improvement	This part is concerned with continuous improvement of the development process and the inspection process.
Rework	The defects identified during the inspection are corrected, and any items requiring investigation are resolved.
Follow up activity	The moderator verifies that the author has made the agreed-upon corrections and completed any investigations.

Table 8.19. Fagan Inspections

The successful implementation of a software inspection program generally has positive impacts on productivity, quality, time to market, and customer satisfaction. For example, IBM Houston employed software inspections for the Space Shuttle missions: 85% of the defects were found by inspections and 15% were found by testing. This project includes about two million lines of computer software. IBM, North Harbor in the United Kingdom quoted a 9% increase in productivity with 93% of defects found by software inspections.

Software inspections have other benefits including their use in educating and training new employees about the product and the standards and procedures to be followed. Sharing knowledge reduces dependencies on key employees. High-quality software has an immediate benefit on productivity, as less time and effort are devoted to reworking the defective product.

The cost of correction of a defect increases the later the defect is identified in the lifecycle. Boehm [Boe:81] states that a requirements defect identified in the field is over forty times more expensive to correct than if it were detected at the requirements phase. It is most economical to detect and fix the defect in the phase that it was introduced. The cost of a requirements defect which is detected in the field includes the cost of correcting the requirements, and the cost of design, coding, unit testing, system testing, regression testing, and so on.

Parnas Approach to Inspections

The use of mathematical documents is at the heart of the approach as the mathematics allows complete coverage. Tabular expressions are employed and these allow a systematic examination of all cases to take place. The Parnas approach is especially valuable in reviewing critical areas of code, where the cost of failure is high, either financially or in loss of life.

The inspection of a document should first identify and list the desired properties of the document. A schedule of inspections is then planned and reviewers with the appropriate background chosen. The reviewers focus on the area of their expertise and they are required to take an active role in the inspection and to justify or reject design decisions. The reviews may focus on checking that the assumptions made are valid, sufficient, and consistent with the function descriptions of each module. The adequacy of the access programs to meet all of the requirements will also be checked.

The reviewers are supplied with a questionnaire to perform the document review. This forces the reviewers to think carefully and to understand the document, as the answers to the questions require a thorough understanding. The questionnaire is designed to describe the properties that the reviewer needs to check and ensures that the reviewer takes an active part in the review. For example, some sample questions in the questionnaire for a review to verify the consistency between assumptions and functions may be:

No.	Questions
1.	Which assumptions tell you that this function can be implemented as described.
2.	Under what conditions may the function be applied?

Table 8.20. Questions at Parnas Inspections

Open-ended questions are employed to force the reviewer to analyze, understand, and make a judgment based upon the facts. The issues raised by the reviewers are discussed in small meetings between the reviewer and designer. The reviewers actively use the code and focus on small sections for the code inspections. The hierarchical structure of the code is exploited rather than proceeding sequentially through the code. The approach is systematic with no case or section of program overlooked. Precise summaries of each short section of code are produced using the tabular expressions. The actual code is then compared with its mathematical description to ensure its correctness.

8.9 Tools

Parnas has been active in developing a set of tools to support his tabular expressions. His SQRL group in Canada developed a suite of tools[20] to support the tabular expressions and Science Foundation Ireland has provided Parnas with €5-6 million of funding to develop a software quality research laboratory (SQRL) at the University of Limerick in Ireland and to develop an enhanced set of tools for the tabular expressions. These tools will include:

- Tabular expression editor
- Printing tabular expressions
- Pre-evaluation checking tool
- Table completeness and consistency checks

---

[20] However, these early tools produced in Canada were a long way from the industrial strength required. The author will be interested in seeing the results from SQRL (www.sqrl.ul.ie) at the University of Limerick.

8.10 Summary

David L. Parnas has been influential in the computing field, and his ideas on software engineering remain important. He advocates a solid engineering approach to the development of high-quality software. Software engineers should apply mathematical and scientific principles to design and develop useful products. He argues that computer scientists should be educated as engineers and should have the right mathematical background to do their work effectively. His contributions include:

- *Tabular Expressions*
Tabular expressions enable complex predicate logic expressions to be represented in a simpler form.

- *Mathematical Documentation*
These include mathematical documents for the system requirements, system design, software requirements, module interface specification, and module internal design.

- *Requirements Specification*
This includes a mathematical model (consisting of relations) to specify the requirements precisely.

- *Software Design*
A module is characterized by its knowledge of a design decision (secret) that it hides from all others. The information hiding principle allows software to be designed for changeability.

- *Software Inspections*
The inspections involve the production of mathematical tables, and may be applied to the actual software or documents.

- *Predicate Logic*
Parnas has introduced an approach to deal with undefined values in predicate logic expressions.

9
Cleanroom and Software Reliability

9.1 Introduction

The Cleanroom approach to software development employs mathematical and statistical techniques to produce high-quality software. The approach was developed by Harlan Mills and others and has been used successfully at IBM. The approach uses mathematical techniques to build quality into the software, and to demonstrate that the program is correct with respect to its specification. Cleanroom also provides certification of the quality of the software based on the expected usage profile of the software.

Software reliability is the probability that the program works without failure for a specified length of time, and is a statement on the future behavior of the software. It is generally expressed in terms of the mean time to failure (MTTF) or the mean time between failure (MTBF).

The release of an unreliable software product at best causes inconvenience to customers, and at worst can result in damage to property or injury to a third party. Consequently, companies need a mechanism to judge the fitness for use of the software product prior to its release and use. The software reliability models are an attempt to predict the future reliability of the software and to thereby assist the project manager in deciding on whether the software is fit for release.

The correction of defects in the software leads to newer versions of the software, and most of the existing reliability models assume reliability growth: i.e., the new version is more reliable than the older version. However, it is essential to gather data to verify this, and some sectors in the safety critical field take the view that the new version is a new program and that no inferences may be drawn until further investigation has been done.

There are many reliability models in the literature and the question as to which is the best model or how to evaluate the effectiveness of the model arises. A good model will have good theoretical foundations and will give useful predictions of the reliability of the software.

9.2 Cleanroom

The objective of the Cleanroom approach is to design software correctly, and to certify the software quality based on the predicated operation usage of the software. The approach was developed by Harlan Mills and others [Mil:87]. The description of Cleanroom presented here is based on [CoM:90]. The key features of Cleanroom include:

Features of Cleanroom
• Specification of statistical usage of software
• Incremental software construction
• Uses rigorous mathematical engineering practices
• Functional verification of software units
• Statistical testing based on usage profiles
• Certification of software quality (MTTF)
• Testing performed independently of development

Table 9.1. Features of Cleanroom

The Cleanroom approach uses development methods that are based on mathematical function theory. The objective is that each function should be complete, consistent, and correct with respect to the requirements. Black box, state box, and clear box development methods are employed for specification and design. The specification is given by a black box description, and this gives an external view of the system in terms of a mapping from the history of the inputs to the outputs of the system. This is then transformed into a state machine view by adding states to reflect the history of inputs, and this is the first step toward implementation. A new state is given by a transition from the current state and input, and the stimulus history no longer needs to be considered, since the current state is a reflection of the input history. The state box is then transformed into a clear box view by adding the procedures that will carry out the state transitions. This may require the introduction of further black boxes for major operations. Each box structure is reviewed to verify its correctness; for example, the correctness of a clear box is demonstrated by showing that it is a valid refinement of the state and black boxes.

Cleanroom follows an iterative and incremental process for software development. Each iteration involves the development of user function software increments, and this eventually leads to the final product. The process includes reviews to build quality into the product and to verify that quality has been built in. These include development reviews and verification reviews. Development reviews focus on technical issues such as the specification, data structures, algorithms, and the emphasis is on identifying and including good ideas. Verification reviews focus on correctness and completeness of work products. The code is reviewed against the specification and design. Any changes to a work product are subject to a subsequent review.

Cleanroom employs mathematical proof to verify the units of software logically rather than using debugging techniques. The Cleanroom approach to unit verification involves showing that the program is a complete rule for a subset of the specification. This is true since a specification is a function or relation whereas a program is a rule for a function. The advantages of the mathematical approach over debugging is:

Cleanroom vs. Debugging
• Design errors are identified earlier. • Eliminates need for unit testing and debugging. • Unit is demonstrated to meet its specification by logical argument. • It takes less time than conventional unit testing and debugging. • Eliminates introduction/finding of subtle defects by debugging.

Table 9.2. Cleanroom vs. Debugging

Functional verification [Mil:79] is the approach used in Cleanroom to verify the correctness of large programs. This involves structuring the proof that a program implements its specification correctly. The program specification is a function (or relation) and the program is a rule for the function. Hence, the proof must show that the rule (program) correctly implements the function (specification). Mills and others applied these techniques in IBM and trained engineers to communicate with other engineers in terms of proofs. They observed that:

Cleanroom Results
• Engineers find the mental challenge of functional verification to be more stimulating than debugging. • Functional verification is well within the capability of engineers (after training). • Many engineers enjoyed the teamwork associated with functional verification. • Improved quality of engineer's work. • Improved productivity of engineer.

Table 9.3. Cleanroom Results

The Cleanroom approach to software testing is quite distinct in that it employs statistical usage testing rather than designing tests that cover every path through the program (coverage testing). The tester draws tests at random from the population of all uses of the software, in accordance with the distribution of expected usage. The tester is required to understand what the software is expected to do and its usage profile. The tester then designs tests that are representative of the expected usage. An execution of the software that does not do what it is required to do is termed an execution failure. Some execution failures are frequent, whereas others are infrequent. Clearly, it is essential to identify any frequent execution failures, as otherwise the users of the software will identify these, resulting in a negative user perception of the software company. One problem with coverage testing is that a tester is equally likely to find a rare execution failure as a frequent execution failure. When a tester discovers an execution failure the software is then analyzed by the software engineers and corrected.

The advantage of usage testing (that matches the actual execution profile of the software) is that it has a better chance of finding the execution failures that occur frequently. The goal of testing is to eliminate frequent execution failures and to maximize the expected mean time to failure. The weakness of coverage testing is evident from the study of nine major IBM products by Adams [Ada:84] as illustrated in the table below:

	Rare					Frequent		
	1	2	3	4	5	6	7	8
MTTF (years)	5,000	1,580	500	158	50	15.8	5	1.58
Avg % failure	33.4	28.2	18.7	10.6	5.2	2.5	1.0	0.4
Probability failure	0.008	0.021	0.044	0.079	0.123	0.187	0.237	0.300

Table 9.4. Adam's Study of Software Failures of IBM Products

The analysis on coverage testing by Adams shows that 61.6% of all fixes (Group 1 and 2) were made for failures that will be observed less than once in 1,580 years of expected use, and that these constitute only 2.9% of the failures observed by typical users. On the other hand, Groups 7 and 8 constitute 53.7% of the failures observed by typical users and only 1.4% of fixes. Therefore, coverage testing is not cost effective in increasing MTTF. Usage testing, in contrast, would allocate 53.7% of the test effort to fixes that will occur 53.7% of the time for a typical user. Mills has calculated that the data in the table show that usage testing is 21 times more effective than coverage testing [CoM:90].

Cleanroom usage testing requires that the expected usage profile of the system be determined. The population of all possible usages is then sampled in accordance with the expected usage profile. The tests are drawn at random from the population of all possible usages in accordance with the expected usage profile. A record of success and failure of the execution of the individual tests is maintained and used to estimate the software reliability.

Cleanroom plays a key role in developing zero-defect software. It allows a probability measure to be associated with the software (based on the predicted operational usage of the software). The software is released only when the probability of zero defects is very high. This involves the use of statistical quality control and analysis of the predicted usage profile of the software. The Software Engineering Institute (SEI) has developed the Cleanroom Reference Model [LiT:96] and this includes management, specification, development and certification processes.

Some of the results on the benefits of using Cleanroom on projects at IBM [CoM:90] are detailed in the table below. These results include quality and productivity measurements.

Project	Results
Flight Control (1987) 33KLOC	Completed ahead of schedule. Error-fix effort reduced by factor of five. 2.5 errors KLOC before any execution
Commercial Product (1988)	Deployment failures of 0.1/KLOC. Certification testing failures 3.4 / KLOC Productivity 740 LOC/month[1]
Satellite Control (1989) 80 KLOC (partial Cleanroom)	50% improvement in quality Certification testing failures of 3.3 / KLOC Productivity 780 LOC/month 80% improvement in productivity
Research Project (1990) 12 KLOC	Certified to 0.9978 with 989 test cases

Table 9.5. Cleanroom Results in IBM

The next section discusses the actual Cleanroom software development process that engineers are required to follow to achieve high-quality software. It is known as Cleanroom software engineering.

Cleanroom Software Development Process

The Cleanroom software development process enables engineers to create high-quality software consistently. There are three teams involved in a Cleanroom project, namely the specification team, the development team, and the certification team. The Cleanroom software development process involves the following phases:

> • *Specification*
> The specification team prepares and maintains the specification and specializes it for each development increment. A rigorous formal specification document is completed and approved prior to commencing design and development. The specification document consists of three parts, namely the external specification, the internal specification and the expected usage profile.

---

[1] Care needs to be taken with productivity measures based upon lines of code.

The external specification describes how the software will look and feel from the user's viewpoint. It includes details on the system environment, system use, performance, etc. It is written in language that users understand.

The internal specification is more mathematical and completely states the mathematical function (or relation) for which the program implements a rule. This definition is required in order to correctly implement and to verify the correctness of the program. However, the internal specification is implementation independent and allows the designers freedom to choose the most appropriate design. The internal specification builds upon the external specification. For example, the external specification defines the stimuli that the software will act upon and the responses produced. The internal specification defines the responses in terms of the stimuli history. The use of stimuli and responses avoids any commitment to a particular implementation thereby allowing the designers greater freedom.

The expected usage profile defines the expected use of the software and guides the preparation of usage tests. The validity of the software's expected MTTF is dependent on tests being run from the population of all executions in the same proportion that they will be generated when the system is in use. The usage model involves the definition of the usage states and estimating the transition probabilities between usage states.

• *Development*
The development team is responsible for designing and implementing the software. There may be more than one development team available and this allows parallel development to take place. The specification is decomposed into small executable increments. Each increment is then designed, implemented, and verified. The increment is tested by invoking user commands for the software. The construction plan for the software will detail the number of increments. The example below is taken from [CoM:90] and shows that Team A is responsible for developing three increments and Team B is responsible for developing one increment. The preparation of test scripts takes place in parallel to the development, and the increments are then certified in parallel with development. The developed increments are integrated and certified and finally all of the increments are integrated are certified.

Team A	Inc 1	Inc 2	Inc 3		
Team B		Inc 4			
Test Scripts	Inc 1	Inc 1,2	Inc 1,2,3	Inc 1,2,3,4	
Certify		Inc 1	Inc 1,2	Inc 1,2,3	Inc 1,2,3,4

Fig. 9.1. Construction Plan for Project

The development team uses box structures, step wise refinement and functional verification to design and develop the software. The development process involves 3 key steps.

Step	Description
1.	Design each increment top down using box structure technology that includes three views of the software and verify the correctness of each view: • *Black Box* Implementation independent view defines responses in terms of stimuli histories. a. Define stimuli. b. Define responses in terms of stimuli histories. • *State Box* The state box is a data-driven view and includes implementation details by modifying the black box to represent responses in terms of current stimuli and the state data that represents stimuli history. a. Define state data to represent stimuli histories. b. Modify black box to represent responses in terms of stimuli and state data. c. Verify state box. • *Clear Box* The clear box is a process driven view and completes implementation details by modifying the state box to represent responses in terms of the current stimuli, state data, and the invocation of lower level black boxes. a. Invent (select) data abstractions to represent state data (e.g., sets, queues, stacks). b. Modify state box to represent responses in terms of stimuli and lower level black boxes. c. Verify the clear box.
2.	Implement each increment by stepwise refinement of clear boxes into executable code. There is a rigorous stepwise refinement algorithm used in Cleanroom for box structures.

3.	Verify that the code meets its specification using functional verification. The proof must show that the rule (program) correctly implements the function (specification). The Cleanroom developers neither compile nor test the program. Instead, mathematical proof (functional verification) is employed. Testing is the responsibility of the certification team.

Table 9.6. Cleanroom Design and Development

• *Certification*
The certification team is responsible for certifying the software's mean time to failure (MTTF) through the application of statistical quality control. The expected usage profile and the applicable parts of the external specification are employed by the certification team to develop test cases for the increment just developed and the increments developed previously. This activity is performed in parallel with development as it uses the specification, not the code. The certification team compiles the increment and adds it to the other increments. The software is certified as follows:

Step	Description
1.	It measures T_k the MTTF of the current version of the software (version k) by executing random test cases. The MTTF for versions 0 ... k-1 have previously been determined. Each test result is compared to the expected result and the cumulative time to failure is an estimate of the MTTF. The failures are reported to development, fixes made, and measurements repeated for new versions of the software.
2.	The reliability of the next version is estimated using a certification model and the measured MTTF for each version of the software. The MTTF is predicted from the formula: $$\text{MTTF}_{k+1} = AB^{k+1}$$ The data points $T_0, \dots T_k$ are fitted to an exponential curve relationship and the values A and B determined. If the value of B is less than 1 then the new version is worse than the previous one. For reliability growth, the value of B should be monotonically increasing.

3.	Once the team has estimated the MTTF for the next version, the team decides whether to: • Correct failures and continue to certify. • Stop certification as desired reliability reached. • Stop certification and redesign the software (the failure rate is too high).

Table 9.7. Cleanroom Certification of Software Reliability

9.3 Statistical Techniques

Probability and statistics were discussed in Chapter 2. The probability of an event occurring is a mathematical indication of the likelihood of the event occurring and the mathematical probability of an event is between 0 and 1. A probability of 0 indicates that the event cannot occur, whereas a probability of 1 indicates that the event is guaranteed to occur. A probability value greater than 0.5 indicates that the event is more likely to occur than not to occur.

A sample space is the set of all possible outcomes of an experiment and an event E is a subset of the sample space. The probability of the union of disjoint events is the sum of their individual probabilities: i.e.,

$$P(\cup^n_{i=1} E_i) = \Sigma^n_{i=1} P(E_i).$$

The probability of the union of two events (not necessarily disjoint) is given by:

$$P(E \cup F) = P(E) + P(F) - P(EF).$$

The probability of an event E not occurring is denoted by $P(E^c)$ and is given by :

$$P(E^c) = 1 - P(E).$$

The probability of an event E occurring given that an event F has occurred is termed conditional probability (denoted by $P(E|F)$) and is given by:

$$P(E|F) = \frac{P(EF)}{P(F)}$$

Bayes formula enables the probability of an event E to be determined by a weighted average of the conditional probability of E given that the event F has occurred and the conditional probability of E given that the event F has not occurred: i.e.,

$$P(E) = P(E|F)P(F) + P(E|F^c)P(F^c).$$

Two events E and F are independent if knowledge that F has occurred does not change the probability that E has occurred. Two events E and F are independent if:

$$P(EF) = P(E)P(F).$$

A good account of probability and statistics is in [Ros:87]. Probability theory has been applied to develop software reliability predictors of the MTTF or the MTBF. Statistical testing is employed extensively in engineering and has been successful in predicting the reliability of hardware. In many situations it is infeasible to test all items in a population so statistical sampling techniques are employed, and the quality of the population is predicted from the quality of the sample. This technique is highly effective in manufacturing environments where variations in the manufacturing process lead to defects in the products. The approach is to take a sample from the batch, which is then used to make judgments as to whether the quality of the batch is acceptable.

Software is different from manufacturing in that the defects are not due to the variations in processes but are due to the design of the software product itself. The software population to be sampled consists of all possible executions of the software. This population is infinite and therefore exhaustive testing is impossible. Statistical testing is used to make inferences on the future performance of the software. Cleanroom usage testing requires that the expected usage profile of the system be determined. The population of all possible usages is then sampled in accordance with the expected usage profile. The tests are drawn at random from the population of all possible usages in accordance with the expected usage profile.

Test cases are generated by traversing the model from the start state to the end state, and then randomly selecting inputs to be included in the test case. Cleanroom testing based on the usage models produces statistically valid inferences about expected operational performance (including MTTF) of a given version of the software [Pro:99].

9.4 Software Reliability

Software has become increasingly important for society, and professional software companies aspire to develop high-quality and reliable software. Many companies desire a mechanism to predict the reliability of their software prior to its deployment and operational use, and this has led to interest in the reliability predictor models that provide an approximate estimate of the reliability of the software. However, accurate predictors of reliability are hard to obtain and there is a need for further research to develop better software reliability models.

DEFINITION (SOFTWARE RELIABILITY)
Software reliability is defined as the probability that the program works without failure for a specified length of time and is a statement on the future behaviour of the software. It is generally expressed in terms of the MTTF or the MTBF. For programs that are only intermittently used, reliability is best defined as the probability that the program works when required.

The reliability of the software is related to the inputs that are entered by the users. Let I_f represent the fault set of inputs (i.e., $i_f \in I_f$ if and only if the input of i_f by the user leads to failure). The randomness of the time to software failure is due to the unpredictability in the selection of an input $i_f \in I_f$. The way in which the user actually uses the software will impact upon the reliability of the software. Software reliability requires knowledge of probability and statistics.

Why Do We Need Software Reliability Models?

The release of an unreliable software product at best causes inconvenience to customers and at worst can result in damage to property or injury to a third party. Consequently, companies need a mechanism to judge the fitness for use of the software product prior to its release and use. The approach taken by most mature companies is to identify a set of objective criteria to be satisfied prior to release. The criteria typically include that all testing has been performed, all tests have passed, all known defects have been corrected, etc. The satisfaction of the criteria provides a degree of confidence that the software has the desired quality and is safe and fit for use. However, the fact that criteria have been satisfied indicates that certain results are true of the software at present or in the past; it may provide no prediction or indication of the future behavior of the software (unless some form of statistical usage testing and reliability prediction is included as part of the criteria). The software reliability models are an attempt to predict the future reliability of the software and to thereby assist the project manager in deciding on whether the software is fit for release.

What Is the Relationship between Defects and Failures?

The relationship between the defect density of a system and failure needs further investigation. A defect may not result in a failure as most software defects are benign. In fact, most observed failures arise from a small proportion of the existing defects. This was demonstrated by research carried out at IBM by Adams [Ada:84] in 1984 on a study of the relationship between defects and failures. The research was based on an analysis of nine large products, and it studied the relationship between detected defects and their manifestations as failures. The results indicate that over 33% of the defects led to an observed failure with MTTF greater than 5000 years; whereas less than 2% of defects led to an observed failure with a MTTF of less than fifty years. Consequently, as a small proportion of defects often give rise to almost all of the observed failures, it is important to identify these as they are the defects that will eventually exhibit themselves as failures to a significant number of users.

Caper Jones [Sta:92] suggests a close relationship between defect density and failures and provides a table of defect densities and the corresponding MTTF. However, the implications of the IBM research (if it is true in general) suggest that it is naive to associate total defects and defect density with software reliability, and that the defect count is a poor predictor of operational reliability. Further, an emphasis on removing a large number of defects from the software is not sufficient in itself to achieve high reliability. The Adams study suggests that it is important to identify and remove the small proportion (2%) of defects with a MTTF of less than fifty years as these will directly impact the observed software reliability. However, the impact of all defects must be considered, as it is unacceptable for defects that may cause major financial loss or injury or loss of life to be allowed to remain in the software. Reliability is complementary to defect density rather than directly related to it.

The study by Adams therefore suggests that care has to be taken with reliability growth models. Clearly, there is no growth in reliability unless the corrected defect (assuming that the defect has been corrected perfectly with no negative impact on the existing software) is a defect that is likely to manifest itself as a failure. Most existing software reliability growth models assume that all remaining defects in the software have an equal probability of failure, and that the perfect correction of a defect leads to an increase in software reliability. The evidence suggests that this premise, on which many of the software reliability growth models are based, is false.

Are All Failures the Same?

Failures vary in their severity and on their impact on the user, and most mature organizations employ a classification scheme that rates the severity of the failure. Many failures are cosmetic and have minimal impact on the system; however, some failures are severe and have a serious adverse impact on the performance of the system. These service-affecting failures require attention as the severity of the problem is such that the customer will be severely financially affected by the failure and may suffer major loss of business (e.g., as happens during a telecommunications outage).

This has led to service level agreements between customers and software companies that guarantee a response to the failures within a stated time period (depending on the severity of the failure). Most service level agreements have built-in penalty clauses that stipulate the compensation that the customer will receive if the customer support does not meet the standards agreed to in the service level agreement.

What Is the Relationship between Testing and Reliability?

Modern software testing is described in [ORg:02] and the reliability of the software is related to the effectiveness of the software testing. Good testing requires good planning; having the appropriate test resources available; a comprehensive suite of test cases to cover the requirements specification; a mechanism to communicate failures to software developers and to track the status of changes made to correct a defect; a mechanism (e.g., defect testing and regression testing) to verify that corrections to failures are properly made and have no adverse impact on the software. Most mature companies plan for testing early in the software development lifecycle and the type of testing performed includes functional, system, performance, and usability testing. Regression testing is employed to verify that the integrity of the system is maintained following the correction of a defect. It aims to verify that the new version of the software is similar in functionality to the previous version. This approach of identifying and correcting defects should therefore (in theory) lead to a growth in reliability of the software and thereby increased confidence in the fitness for use of the software. However, care is needed before drawing this conclusion (as discussed earlier) as the identified defects may be in parts of the software that are infrequently used.

The Cleanroom approach employs statistical usage testing rather than coverage testing and applies statistical quality control to certify the MTTF of the software. The statistical usage testing involves executing tests chosen from the population of all possible uses of the software in accordance with the probability of expected use. This requires an understanding of expected usage of the software and what it is intended to do. Conventional testing is as likely to find a rare execution failure as well as a frequent execution failure; however, what is required is to find failures that occur on frequently used parts of the system. The Cleanroom approach (as distinct from the defect count approaches) offers a meaningful approach to software reliability.

Fenton [Fen:00] has investigated the relationship between pre- and post release defects. His results suggest that modules which are the most defect prone pre release are among the least defect prone post release and vice versa. Further empirical studies are needed to see whether these results are generally true.[2]

How Are the Old and New Versions of Software Related?

Software failures that are identified are corrected in the new version of the software. The precise relationship between the new version of the software and the previous version needs to be considered. The different views include:

---

[2] My experience is that the maturity of the test group and the test practices employed need to be taken into account before making a judgment on this. A software module that is not error prone pre release and is error prone post release may have been tested inadequately by the test group.

Similarities and Differences between New/Old Version
• The new version of the software is identical to the previous version except that the identified defects have been corrected.
• The new version of the software is identical to the previous version except that the identified defects have been corrected but there may be some new defects introduced by developers during the fix process.
• No assumptions can be made about the behavior of the new version of the software until further data are obtained.[3]

Table 9.8. New and Old Version of Software

Many of the existing software reliability models assume reliability growth: i.e., the new version is assumed to be more reliable than the previous version, as several identified defects have been corrected. It is essential that a new version be subject to rigorous regression testing to have confidence that its behavior is similar to the previous version, and that no new defects have been introduced that would degrade the reliability of the software. In the absence of comprehensive regression testing, no assumptions should be made between the relationship of the new version of the software to the previous version.

The safety critical industry (e.g., the nuclear power industry) takes the conservative viewpoint that any change to a program creates a new program. The new program is therefore required to demonstrate its reliability again. That is, data need to be gathered on the new program and no assumptions can be made about its reliability until the studies are complete.

What Reliability Models are Currently Used?

There are many well-known predictor models employed (with varying degrees of success), and a description and critical examination of some of these is available in [BeM:86]. Some of these are not reliability models according to the strict definition of reliability that was discussed earlier, and several of them just compute defect counts rather than providing an estimate of software reliability in terms of MTTF. However, the objective here is to describe a selection of what is currently employed in the literature. The fact that there is such a wide collection of models suggests that there is no one model at this time that serves the needs of the software engineering community.

---

[3] It is essential to perform regression testing to provide confidence that the new version of the software maintains the integrity of the software.

Model	Description	Comments
Jelinski / Moranda Model	The failure rate is a Poisson process and is proportional to the current defect content of program. The initial defect count is N; the initial failure rate is $N\varphi$; it decreases to $(N-1)\varphi$ after the first fault is detected and eliminated, and so on. The constant φ is termed the proportionality constant.	Assumes defects corrected perfectly and no new defects are introduced. Assumes each fault contributes the same amount to failure rate.
Littlewood/ Verrall Model	Successive execution time between failure independent exponentially distributed random variables. Software failures are the result of the particular inputs and faults introduced from the correction of defects.	Does not assume perfect correction of defects.
Musa Execution Time Model	Refinement of Jelinski/Moranda Model. Fault correction rate proportional to rather than equal to the failure rate.	Assumes each defect contributes same amount to overall failure rate.
Seeding and Tagging	This is analogous to estimating the fish population of a lake. One approach (Mills) is to introduce a known number of defects into a software program and to monitor the proportion of inserted defects identified during testing. Another approach (Hyman) is to regard the defects found by one tester as tagged and then to determine the proportion of tagged defects found by a second independent tester.	Provides an estimate of the total number of defects in the software. Does not enable reliability to be determined. Assumes all faults are equally likely to be found. Assumes the introduced faults are representative of existing.
Generalized Poisson Model	Number of failures observed in i^{th} time interval τ_i has a Poisson distribution with mean $\phi(N-M_{i-1})\,\tau_i^{\alpha}$ where N is the initial number of faults; M_{i-1} is the total number of faults removed up to the end of the $(i-1)^{th}$ time interval; ϕ is the proportionality constant.	Assumes faults removed perfectly at end of time interval.
Time Series Analysis	This is a technique to analyze variable data over time. It allows data to be analyzed without any assumptions about the failure process.	

Table 9.9. Software Reliability Models

The existing reliability models in the literature may be classified into the following:[4]

• *Size and Complexity Metrics*
These are used to predict the number of defects that a system will reveal in operation or testing. Many of these are regression based "data fitting" models and includes work by Halstead, Fenton, Kitchenham, and McCabe.

• *Operational Usage Profile*
These are used to predict failure rates based on the expected operational usage profile of the system. The approach taken is to look at the product and test it. The number of failures encountered is determined and the reliability is then predicted. The Cleanroom approach developed by Mills at IBM employs statistical usage testing to predict the MTTF.

• *Quality of the Development Process*
These are used to predict failure rates based on the maturity of the organization or of the software development process. The CMM [Pau:93], and its successor, the CMMI [CKS:03], are examples of maturity models. Companies that have been assessed at level three or higher are expected to deliver high quality software.[5]

What Is a Good Software Reliability Model?

Models are simplifications of the underlying reality and enable predictions about future behavior to be made. A model is a foundation stone from which the theory is built, and from which explanations and justification of behavior are made. The model is in effect the starting point and it is not envisaged that we should justify the model itself. However, if the model explains the known behavior of the system, it is thus deemed adequate and a suitable representation.

The adequacy of the model is thus a key concern, and in normal science [Kuh:70] inadequate models are replaced with more accurate models. The adequacy of the model is judged from model exploration, and empirical analysis is used to determine if the predictions of the model are close to the actual manifested behavior. If empirical evidence supports the model, then the model is judged to be a good representation of the underlying reality. However, if serious inadequacies are identified with the model, then the theory and its foundations collapse like a house of cards. In some cases it may be possible to amend the model to address its inadequacies. In practice, models are modified or replaced over time, as further facts and observations lead to aberrations that cannot be explained by the model in its current form.

---

[4] Some of these are not software reliability models according to the definition of software reliability as the MTTF.

[5] I have seen instances where maturity of processes did not lead to higher quality software and on-time projects. This included a project delivered by a CMM level 5 company in India.

The physical world is dominated by mathematical models: e.g., models of the weather system, that enable predictions of the weather to be made. The extent to which the model explains the underlying physical behavior and allows predictions of future behavior to be made will determine its acceptability as a representation of the physical world. Models are usually good at explaining some aspects of the world and weak at explaining other aspects. There are many examples in science of the replacement of one theory by a newer one: e.g., the replacement of the Ptolemaic model of the universe by the Copernican model or the replacement of Newtonian physics by Einstein's theories on relativity.

There are many well-known software reliability models but their ability to predict failure accurately has had limited success. Many prediction models tend to model only part of the underlying problem. A good software reliability model will have the following characteristics:

Characteristics of Good Software Reliability Model
• Good theoretical foundation
• Realistic assumptions
• Valid empirical support
• As simple as possible (Occam's Razor)
• Trustworthy and accurate

Table 9.10. Good Software Reliability Model

It is essential that the evidence for a proposed software reliability model is valid.

How Is a Software Reliability Model Evaluated?

The extent to which the software reliability model can be trusted depends on the accuracy of its predictions and on its soundness. Empirical data will need to be gathered to determine the extent to which the observations support the predictions of the model. It may be acceptable to have a little inaccuracy in the predictions provided the predictions are close to the observations. A model that gives overly optimistic results is termed *optimistic*, whereas a model that gives overly pessimistic results is termed *pessimistic*. Inaccuracy may be acceptable in the early stages of prediction, provided that when the model is employed to predict operational reliability, it is accurate. The assumptions inherent in the reliability model need to be examined to determine whether they are realistic. Several well-known software reliability models include some of the following questionable assumptions:

- All defects are corrected perfectly.
- Defects are independent of one another.
- Failure rate decreases as defects are corrected.
- Each fault contributes the same amount to the failure rate.

The validity of the assumptions made will determine whether the model is sound, useful, and trustworthy.

How Are Software and Hardware Reliability Related?

There are similarities and differences between hardware and software reliability. Hardware failure is often due to a component wearing out due to its age, and in most cases a hardware failure is permanent and requires replacement of a hardware component. Most hardware components are expected to last for a certain period of time, and the variation in the failure rate of a hardware component is due to the manufacturing process and to the operating environment of the component. Good hardware reliability predictors have been developed, and each hardware component has an expected MTTF. The reliability of a product may be determined from the reliability of the individual components of the product.

Software is different in that it does not physically wear out and instead software failures are the result of particular inputs. There is no variation in manufactured software as each copy is identical. Software is either correct or incorrect, and software failures are due to design and implementation errors. The software community has been ineffective in developing a sound predictor model of software reliability. Several models have been discussed, but these contain questionable assumptions that limit their usefulness as a predictor of software reliability.

How Are Software Availability and Reliability Related?

Software availability is a measure of the down-time of the software during a particular time period. The down-time refers to a period of time when the software is unavailable for use and a goal of approximately five minutes downtime per annum (known as five nines) is common in the telecommunications sector. A company with a goal of five-nines availability aims to develop software that is available for use 99.999% of the time in the year.

Software that satisfies strict availability constraints is usually reliable. The downtime includes the time in rebooting a machine, upgrading to a new version of software, planned and unplanned outages. It is theoretically possible for software to be highly unreliable but yet satisfy the five-nines goal. Consider, for example, software that fails consistently for 0.5 seconds every day. Then the total failure time is 183 seconds or approximately 3 minutes. However, this scenario is unlikely for most systems, and the satisfaction of strict availability constraints means that the software is highly reliable.

Software that is highly reliable may satisfy poor availability constraints. Consider the upgrade of the version of software at a customer site to a new version. Suppose the previous version had been highly reliable but that the upgrade path is complex or poorly designed (e.g., taking two days). Then the availability measure is very poor even though the product is highly reliable. Consequently, care is required before drawing conclusions between software reliability and software availability.

Why Are There So Many Reliability Models?

The large collection of reliability models in the literature suggest that little progress has been made in the definition of a sound reliability model. It suggests that there is little consensus among the software reliability community as to what is the best model among the plethora of existing models. There is no one model at this time that is able to meet the needs of software practitioners, and further research on a more accurate software reliability predictor is required.

Many of the existing models lack a solid theoretical foundation and adopt a graph fitting approach to prediction. However, what is required is a sound software reliability model with reasonable assumptions that gives accurate predictions.

Operational Profile and Software Reliability

The way in which the system is used will impact the quality and reliability as perceived by the individual user. Failures will manifest themselves on certain input sequences only, and as users generally employ different input sequences, so each user will have a different perception of the reliability of the software. The knowledge of the way that the software will be used allows the software testing to focus on verifying that the software works correctly for the normal everyday tasks carried out by users.

Therefore, it is important that the operational profile of users be determined to allow effective testing of the software to take place. The operational environment may not be stable as users may potentially change their behavior over time. The collection of operational data involves identifying the operations to be performed and the probability of that operation being performed. The Cleanroom approach [CoM:90] applies statistical techniques to enable a software reliability measure to be calculated based upon the expected usage of the software.

How Can Software Design Improve Reliability?

Good design is essential in building a high-quality and reliable product. Engineers design bridges prior to their construction and their approach is to first specify the requirements and then to produce a design that will satisfy the requirements. It is important to engineer software and to precisely state the requirements and then to produce a design that will satisfy the requirements. The objective is to build quality and reliability into the software by using sound design methods and tools.

Parnas and his research group adopts a classical engineering approach to improve design. This involves the use of mathematical documents that allow the software requirements to be expressed precisely using tabular expressions. The approach followed by Parnas for specification and design includes the ideas of modularity, information hiding, and the 4-variable model.

Other groups in industry are improving their software design by using the Unified Modeling Language (UML) for requirements and design.

How Can Software Development Tools Improve Reliability?

Various methods and tools may be employed to improve software reliability. A good software development infrastructure is essential and needs to include tools for configuration management, code coverage, test automation, etc. Automated testing tools for regression testing are invaluable as they allow the new version of the software to be efficiently tested and to provide confidence that the quality of the new version is superior to the previous version.

Various research tools are being developed by academic groups, for example, tools to support tabular expressions to aid the specification and design of software, and also to assist inspections and testing. The objective is to develop methods and tools that will provide tangible improvement in software reliability and enhanced confidence in the correctness of the software.

What Can Be Learned from the Reliability of a Project?

It is unfortunate that many projects often repeat the same mistakes of previous projects. The defect and failure profile of the project offers the project team the potential to learn lessons from the project and to do things differently on the next project. The phase of origin of the failure (e.g., requirements, design, coding, test, postrelease, etc.); the cause of nondetection of the failure during inspections and testing; the modules in which the failures occurred; and so forth need to be analyzed and actions identified to prevent a reoccurrence of the defects. This should lead to practical improvement in the next project (e.g., improvements to the design methodology, inspections and testing, etc). It is as important to learn lessons from failure as it is to repeat current success.

9.5 Summary

Software has become increasingly important for society and professional software companies aspire to develop high-quality and reliable software. This has led to an interest in methodologies that have been demonstrated to yield superior results in achieving quality software. Cleanroom is one such approach and employs mathematical and statistical techniques to produce high-quality software. The approach was developed by Harlan Mills and others and has been used successfully at IBM. The approach builds quality into the software and uses mathematical techniques to demonstrate that the program is correct with respect to its specification. Cleanroom also provides certification of the quality of the software based on the expected usage profile of the software.

Software reliability is the probability that the program works without failure for a specified length of time, and it is a statement on the future behavoiur of the software. It is generally expressed in terms of the mean time to failure (MTTF) or the mean time between failure (MTBF), and the software reliability measurements are an attempt to predict the future reliability of the software and to thereby allow an objective judgment of the fitness for use of the software.

There are many reliability models in the literature and the question as to which is the best model or how to evaluate the effectiveness of the model arises. A good model will have good theoretical foundations and will give useful predictions of the reliability of the software.

10
Unified Modeling Language

10.1 Introduction

The unified modeling language (UML) is a visual modeling language for software systems. It was developed by Jim Rumbaugh, Grady Booch, and Ivar Jacobson [Jac:99a] as a notation for modeling object-oriented systems. It provides a visual means of specifying, constructing, and documenting the object-oriented system, and facilitates the understanding of the architecture of the system and the management of the complexity of a large system. The unified modeling language was strongly influenced by three methods: the Object Modeling Technique (OMT) developed by Rumbaught; the Booch method; and the Object-Oriented Software Engineering (OOSE) developed by Jacobson. It unifies and improves upon these methods and was developed at the Rational Corporation.[1] Although, it is not a mathematical approach to software quality, it is a formal approach to modeling software systems that has become popular,[2] and it is therefore included in this book.

An engineer will design a house prior to its construction and the blueprints will contain details of the plan of each room in the house as well as the practical details of electricity and plumbing. These plans form the basis for the estimates of the time and materials required to construct the house. UML is useful in modeling the system, and a model simplifies the underlying reality. Models provide a better understanding of the system to be developed, and a UML model allows the system to be visualized prior to implementation. Large complex systems are difficult to understand in their entirety, and the models simplify the complexity.

Models simplify the reality, but it is important to ensure that the simplification does not exclude any important details. The chosen model affects the view of the system, and different roles require different viewpoints of the proposed system. A database developer will focus on entity-relationship models, whereas a systems analyst will focus on algorithmic models. An object-oriented developer will focus on classes and interactions of classes. Often, there is a need

---

[1] The Rational Corporation is now part of IBM.

[2] UML is the most widely used formal method in industry. It has good tool support.

to be able to view the system at different levels of detail. No single model in itself is sufficient for this, and a small set of interrelated models is employed. For example, in the plans for a house there are floor plans, electrical plans, and plumping plans.

UML is applied to formally model the system and it allows the same information to be presented in many different ways and at different levels of detail. The requirements of the system are expressed in terms of use cases; the design view captures the problem space and solution space; the process view models the systems processes; the implementation view addresses the implementation of the system; and a deployment view. There are several diagrams providing different viewpoints of the system. The UML diagrams provide the blueprint of software and are discussed later in this chapter.

10.2 Overview of UML

UML is a very expressive graphical modeling language for visualizing, specifying, constructing, and documenting a software system. It provides several views of the software's architecture that are needed to develop and deploy systems. There is a clearly defined syntax and semantics[3] for every building block of the graphical notation of UML. Each stakeholder (e.g., project manager, developers, testers) has a different perspective and looks at the system in different ways at different times over the project's life. UML is a way to model the software system before implementing it in a programming language, and the explicit model of the system facilitates communication.

A UML specification involves building precise, complete, and unambiguous models. The UML models may be employed to generate code in a programming language such as Java or C++. The reverse is also possible and it is therefore possible to work in the graphical notation of UML or the textual notation of a programming language. Tools are employed to keep both views consistent. UML expresses things that are best expressed graphically, whereas a programming language expresses things that are best expressed textually. UML may be employed to document the software system, and it has been employed in many domains including the banking sector, defense, and telecommunications.

The application of UML requires an understanding of the basic building blocks of UML, the rules for combining the building blocks, and the common mechanisms that apply throughout the language. There are three kinds of building blocks employed namely things, relationships and diagrams.

Things are the object-oriented building blocks of the UML. They include structural things, behavioral things, grouping things, and annotational things. Structural things are the nouns of the UML models; behavioral things are the dynamic parts of the UML models and represent behavior over time;

---

[3] Parnas has joked that UML would be better described as the 'Undefined Modeling Language'. However, while I accept that there is work to do on the UML semantics it is nevertheless the most acceptable formal approach to software development in industry and is likely to remain so.

grouping things are the organization parts of UML; and annotation things are the explanatory parts of UML. Things, relationships, and diagrams are all described graphically and are described in detail in [Jac:99a].

Thing	Kind	Description
Structural	Class	A class is a description of a set of objects that share the same attributes and operations.
	Interface	An interface is a collection of operations that specify a service of a class or component. It describes externally visible behavior of the element. It is a specification rather than an implementation.
	Collaboration	A collaboration has structural and behavioral dimensions. It defines an interaction and includes roles.
	Use case	A use case is a description of a set of sequences of actions that a system performs. It yields a result to a particular actor.
	Active class	This is similar to a class except that its objects represent elements whose behaviour is concurrent with other elements.
	Component	A component is a physical and replaceable part of a system that conforms to and realizes a set of interfaces.
	Node	A node is a physical element that exists at run time and represents a computational resource with processing capability and memory.
Behavioural	Interaction	This comprises a set of messages exchanged among a set of objects.
	State cachine	A state machine is a behavior that specifies the sequences of states that an object or an interaction goes through during its lifetime in response to events.
Grouping	Packages	These are the organization parts of UML models. A package organizes elements into groups and is a way to organize a UML model.
Annotation		These are the explanatory parts of UML.

Table 10.1. Classification of UML Things

There are four kinds of relationship in UML namely dependency, association, generalization and realization.

Dependency is a semantic relationship between two things in which a change to one thing affects the other thing (dependent thing). Dependencies show one thing using another. An association is a structural relationship that describes a set of links (connections among objects). Aggregation is an association that represents a structural relationship between a whole and its parts. A generalization is a specialization/generalization relationship in which the objects of the specialized element (child) are substituted for objects of the generalized element (the parent). The child shares the structure and behavior of the parent.

Realization is a semantic relationship between classifiers, where one classifier specifies a contract that another classifier guarantees to carry out. They are encountered in interfaces (classes and components that realize them) and use cases (collaborations that realize them).

The UML diagrams provide a graphical visualization of the system from different viewpoints. A diagram may contain any combination of things and relationships. There are several UML diagrams employed and they include:

Diagram	Description
Class	This shows a set of classes, interfaces, and collaborations and their relationships. They address the static design view of a system.
Object	This shows a set of objects and their relationships. They represent the static design view of the system but from the perspective of real cases.
Use Case	These show a set of use cases and actors and their relationships. They are important in modeling the behavior of a system.
Sequence	These are interaction diagrams that show an interaction of a set of objects and their relationships including messages dispatched between them. A sequence diagram emphasizes the time ordering of messages.
Collaboration	A collaboration diagram is an interaction diagram that emphasizes the structural organization of objects that send and receive messages.
Statechart	This shows a state machine consisting of states, transitions, events, and activities. It addresses the dynamic view of a system and is important in modeling the behavior of an interface or class.
Activity	This is a kind of statechart diagram that shows the flow from activity to activity of a system. It addresses the dynamic view of a system and is important in modeling the function and flow of control among objects.
Component	This show the organizations and dependencies among components. It addresses the static implementation view of a system.
Deployment	This shows the configuration of run time processing nodes and the components that live on them.

Table 10.2. UML Diagrams

UML is often used as part of the unified software development process. The unified process is described in detail in [Jac:99b]. It is:

- Use-case driven
- Architecture centric
- Iterative and incremental

It includes cycles, phases, workflows, risk mitigation, quality control, project management, and configuration control. Software projects may be very complex, and there are risks that requirements may be missed in the process, or that the interpretation of a requirement may differ between the customer and developer. Requirements are gathered as use cases in the unified process, and the use cases describe the functional requirements from the point of view of the users of the system.

The use case model describes what the system will do at a high-level, and there is user focus in defining the scope the project. Use cases drive the development process and the developers create a series of design and implementation models that realize the use cases. The developers review each successive model for conformance to the use-case model. The testers test the implementation to ensure that the implementation model correctly implements the use cases.

10.3 UML Diagrams

The UML diagrams are useful for visualizing, specifying, constructing, and documenting the software architecture. This section provides a more detailed account of the diagrams, and the first diagram considered is the class diagram. Classes are the most important building block of any object-oriented system, and a class is a set of objects that share the same attributes, operations, relationships, and semantics [Jac:99a].

Class diagrams are a superset of the entity-relationship diagrams that are used for logical database design. Classes may represent software things and hardware things. In a house things like walls, doors, and windows are all classes, whereas individual doors and windows are objects. A class represents a set of objects rather than an individual object. UML provides a graphical representation of a class.

Automated bank teller machines (ATMs) include two key classes: customers and accounts. The class definition includes the data structure for customers and accounts and also the operations on customers and accounts. These include operations to add or remove a customer, and operations to debit or credit an account or to transfer from one account to another. There are several instances of customers and accounts, and these are the actual customers of the bank and their accounts.

Customer	Account
Name: String Address: String	Balance: Real Type: String
Add() Remove()	Debit() Credit() CheckBal() Transfer()

Table 10.3. Simple Class Diagram

Every class has a name (e.g., Customer and Account) to distinguish it from other classes. There will generally be several objects associated with the class. The class diagram describes the name of the class, its attributes, and its operations. An attribute represents some property of the class that is shared by all objects; e.g., the attributes of the class Customer are name and address. Attributes are listed below the class name in the UML diagram. The operations are listed below the attributes. The operations may be applied to any object in the class. The responsibilities of a class may also be included in the definition. The concept of class and objects are taken from object-oriented design.

Class diagrams typically include various relationships between classes. In practice, very few classes stand alone, and most collaborate with others in various ways. The relationship between classes needs to be considered and these provide different ways of combining classes to form new classes. The relationships include dependencies (a change to one thing affects the dependent thing); generalizations (these link generalized classes to their specializations in a subclass/superclass relationship); and associations (these represent structural relationships among objects).

A dependency is a relationship that states that a change in the specification of one thing affects the dependent thing. It is indicated by a dashed directed line (---->). Dependencies show one thing using another. Generalizations allow a child class to be created from one or more parent classes (single or multiple inheritance). A class that has no parents is termed a base class. This is illustrated by the example in [Jac:99a] in which the base class is Shape and there are three children of the base class namely Rectangle, Circle and Polygon. There is one child of Rectangle namely Square. Generalization is indicated by a solid directed line that points to the parent (—▶). Association is a structural relationship that specifies that objects of one thing are connected to objects of another thing.

The next diagram considered is the object diagram which shows a set of objects and their relationships at a point of time. The object diagram is related to the class diagram in that the object is an instance of the class. The ATM example above had two classes (customers and accounts) and the objects of these classes are the actual customers and their corresponding accounts. Each customer may have several accounts. The names and addresses of the customers are detailed as well as the corresponding balance in the customer's accounts. There is one instance of the customer class below and two instances of the account class:

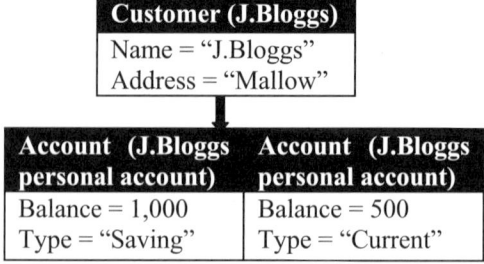

Fig. 10.1. Simple Object Diagram

An object has a state and the value of its state is given at a particular moment of time. Operations on the object typically change its state with the exceptions of operations that query the state. Object diagrams give a snapshot of the system at a particular moment of time. An object diagram contains objects and links to other objects.

The next diagram considered is the use-case diagram. Use cases diagrams model the dynamic aspects of the system and are important in visualizing and specifying the behavior of the system. A use-case diagram shows a set of use cases and actors and their relationships. They describe scenarios or sequences of actions for the system from the user's viewpoint (actor), and show how the actor interacts with the system. An actor represents the set of roles that a user can play while interacting with the system, and an actor may be human or an automated system. Actors are connected to use cases by association and they may communicate by sending and receiving messages.

A use-case diagram shows a set of use cases and each use case represents a functional requirement. Use cases are employed to model the visible services that the system provides within the context of its environment and for specifying the requirements of the system as a black box. Each use case carries out some work that is of value to the actor, and the behavior of the use case is described by the flow of events in text. The description includes the main flow of events for the use case and the exceptional flow of events. These flows may also be represented graphically. There may also be alternate flows as well as the main flow of the use case. Each sequence is termed a scenario and a scenario is one instance of a use case.

Use cases provide a way for the end users and developers to share a common understanding of the system. They may be applied to all or part of the system (subsystem), and the use cases are used as the basis for development and testing. A use case is represented graphically by an ellipse. The benefits of use cases include:

- Enables domain experts, developers, testers, and end users to share a common understanding.
- Models the requirements of the system (specifies what the system should do).
- Models the context of a system (identifying actors and their roles).
- Serves as a basis for development and testing.

A simple example is the definition of the use cases for an ATM application. The typical user operations at an ATM machine include the balance inquiry operation, cash withdrawal, and the transfer of funds from one account to another. The actors for the system are identified, and for the ATM example the actors "customer" and "admin" are employed. These actors have different needs and expectations of the system.

The behavior from the user's viewpoint is described in the following diagram and the use-cases include "withdraw cash," "check balance," "transfer," and "maintain/reports." The use case view of the system includes the actors who are performing the sequence of actions.

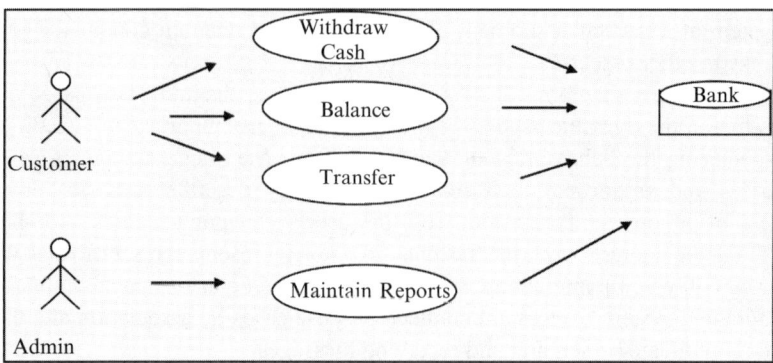

Fig. 10.2. Use-Case Diagram of ATM Machine

The next UML diagram considered is the sequence diagram which models the dynamic aspects of the system and shows the interaction between objects/classes in the system for each use case. The interactions model the flow of control that characterizes the behavior of the system, and a sequence diagram emphasizes the time ordering of messages. The objects that play a role in the interaction are identified. The interactions may include messages that are dispatched from object to object, and messages are ordered in sequence by time. The example as adapted from [CSE:00] considers the sequences of interactions between objects for the "check balance" use case. This sequence diagram is specific to the case of a valid inquiry, and there are generally sequence diagrams to handle the exception cases also.

The behavior of the "check balance" operation is evident from the diagram. The customer inserts the card into the ATM machine and the PIN number is requested by the ATM machine. The customer then enters the number and the ATM machine contacts the bank for verification of the number. The bank confirms the validity of the number and the customer then selects the balance enquiry. The ATM contacts the bank to request the balance of the particular account and the bank sends the details to the ATM machine. The balance is displayed on the screen of the ATM machine. The customer then withdraws the card. The actual sequence of interactions is evident from the sequence diagram.

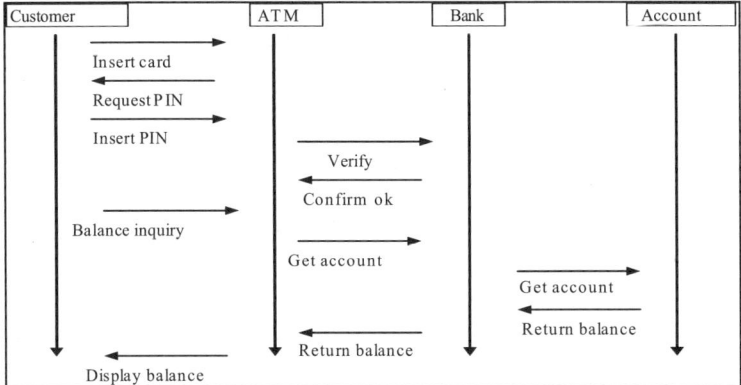

Fig. 10.3. UML Sequence Diagram

The example above has four objects (Customer, ATM, Bank, and Account) and these are laid out from left to right at the top of the sequence diagram. Collaboration diagrams are interaction diagrams that consist of objects and their relationships. However, while sequence diagrams emphasize the time ordering of messages, a collaboration diagram emphasizes the structural organization of the objects that send and receive messages. Sequence diagrams and collaboration diagrams are semantically equivalent to one another and one can be converted to the other without loss of information. Collaboration diagrams are described in more detail in [Jac:99a].

The next UML diagram considered is the activity diagram which are essentially charts showing the flow of control from one activity to another. They are used to model the dynamic aspects of a system, and this involves modeling the sequential and possibly concurrent steps in a computational process. They are different from sequence diagrams in that they show the flow from activity to activity, whereas sequence diagrams show the flow from object to object.

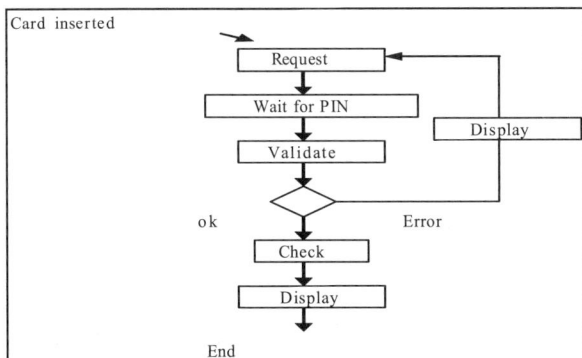

Fig. 10.4. UML Activity Diagram

The final UML diagram discussed is state diagrams or state charts. These show the dynamic behavior of a class and how different operations result in a change of state. There is an initial state and a final state and the different operations result in different states being entered and exited.

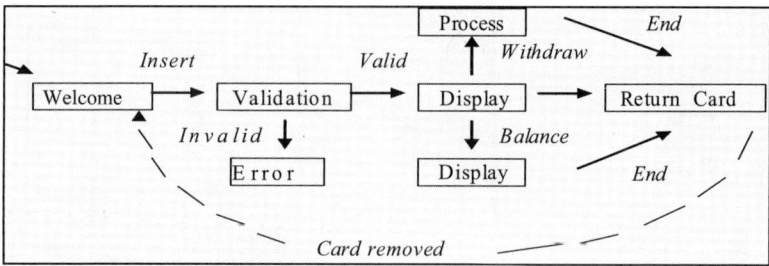

Fig. 10.5. UML State Diagram

There are several other UML diagrams including component and deployment diagrams. The reader is referred to [Jac:99a].

10.4 Object Constraint Language

The object constraint language (OCL) is a specification language that allows a precise way of expressing constraints on the UML models. OCL is a part of UML, although it was originally developed by Jos Warmer at IBM as a business modeling language. It was developed further by the Object Management Group (OMG) as part of the development of UML.[4] OCL is a pure expression language: i.e., there are no side-effects as in imperative programming languages. Expressions can be used in various places in a UML model including:

- specify the initial value of an attribute.
- specify the body of an operation.
- specify a condition

There are several types of OCL constraints including:

OCL Constraint	Description
Invariant	A condition that must always be true. An invariants may be placed on an attribute in a class and this has the affect of restricting the value of the attribute. All instances of the class are required to satisfy the invariant. An

---

[4] I am aware of some groups in the United States that are using OCL to specify constraints. However, many other groups are employing natural language to specify constraints at this stage.

	invariant is a predicate and is introduced after the keyword **inv**.
Precondition	A condition that must be true before the operation is executed. A precondition is a predicate and is introduced after the keyword **pre**.
Postcondition	A condition that must be true when the operation has just completed execution. A postcondition is a predicate and is introduced after the keyword **post**.
Guard	A condition that must be true before the state transition occurs.

Table 10.4. OCL Constraints

There are various tools available to support OCL and these include OCL Compilers (or Checkers) that provide syntax and consistency checking of the OCL constraints and the USE tool which may be employed to animate the model.

10.5 Rational Unified Process

Software projects with a complex problem to solve need a well-structured process to achieve its results. This section gives a brief introduction to the Unified Development Software Process, and to some of its core workflows. The unified process uses the visual modeling standard of UML and a full description of the process is in [Jac:99b]. It is use case driven, architecture centric and iterative and incremental.

The unified process includes cycles, phases, workflows, risk mitigation, quality control, project management, and configuration control. Software projects may be very complex, and there are risks that requirements may be missed in the process, or that the interpretation of a requirement may differ between the customer and developer. Requirements are gathered as use cases in the unified process, and the use cases describe the functional requirements from the point of view of the user of the system. The use case model describes what the system will do at a high level, and there is a user focus in defining the scope of the project. Use cases drive the development process and the developers create a series of design and implementation models that realize the use cases. The developers review each successive model for conformance to the use-case model. The testers test the implementation to ensure that the implementation model correctly implements the use cases.

The software architecture concept embodies the most significant static and dynamic aspects of the system. The architecture grows out of the use cases and factors such as the platform that the software is to run on, deployment considerations, legacy systems, and nonfunctional requirements.

A commercial software product is a large undertaking that may involve 50 to 100 person-years. It may take a year or longer to complete and the work is decomposed into smaller slices or mini-projects, where each mini-project is an iteration that results in an increment.

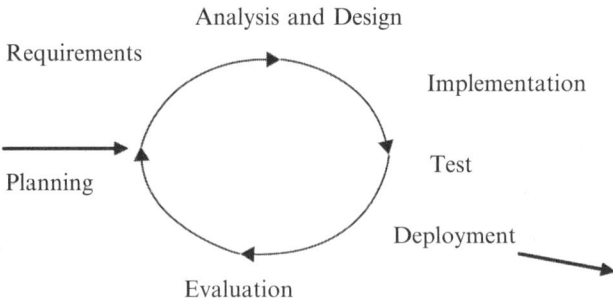

Fig. 10.6. Rational Unified Process

Iterations refer to steps in the workflow, and an increment leads to the growth of the product. The iterations are controlled in the unified process. Controlled iteration reduces the cost risk to the expenditures on a single increment.

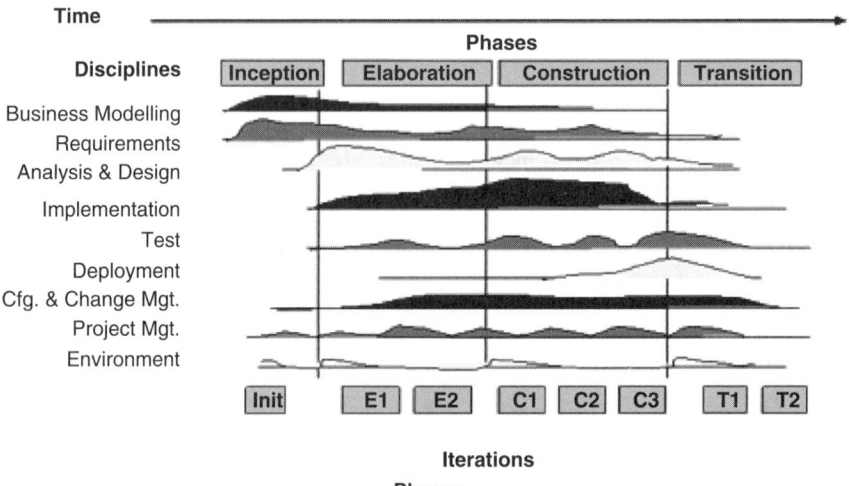

Fig. 10.7. Phases and Workflows in Unified Process

If the developers need to repeat the iteration, the organization loses only the misdirected effort of one iteration, rather than the entire product. Therefore, the unified process is a way to reduce risk in software engineering.

The waterfall software development model is well known and is used frequently in practice. However, it has the disadvantage that the risk is greater toward the end of the project, where it is costly to undo mistakes from earlier phases. Iterative processes were developed in response to these waterfall characteristics. With an iterative process, the waterfall steps are applied iteratively. Instead of developing the whole system in one step, an increment (i.e., a subset of the system functionality) is selected and developed, then another increment, and so on. The earliest iterations address greatest risks. Each iteration produces an executable release and includes integration and testing activities.

The unified process consists of four phases. These are inception, elaboration, construction, and transition. Each phase consists of one or more iterations, and the iteration consists of various workflows. The workflows may be requirements, analysis, design, implementation, and test. Each phase terminates in a milestone with one or more project deliverables.

The inception identifies and prioritizes the most important risks. The phase is concerned with the initial project planning and cost estimation and initial work on the architecture and functional requirements for the product. The elaboration phase specifies most of the use cases in detail and the system architecture is designed. The construction phase is concerned with building the product. At the end of this phase, the product contains all of the use cases that management and the customer agreed for the release. The system is ready for transfer to the user community. The transition phase covers the period during which the product moves into the customer site and includes activities such as training customer personnel, providing help-line assistance, and correcting defects found after delivery.

10.6 Tools for UML

There are many tools available for UML and the following is a selection of the IBM/Rational tools:

UML Tool	Description
Requisite Pro	Requirements and use case management tool. It provides requirements management and traceability.
Rational Software Modeler (RSM)	Visual modeling and design tool that is used by systems architects/systems analysts to communicate processes, flows, and designs.
Rational Software Architect (RSA)	RSA is a tool that enables good architectures to be created.
Clearcase/ Clearquest	These are configuration management/change control tools that are used to manage change in the project.

Table 10.5. Tools for UML

10.7 Summary

The unified modeling language is a visual modeling language for software systems, and it facilitates the understanding of the architecture of the system and the management of the complexity of large systems. It was developed as a notation for modeling object-oriented systems, and it provides a visual means of specifying, constructing, and documenting the object-oriented system.

UML is applied to formally model the system and it allows the same information to be presented in many different ways and at different levels of detail. The requirements of the system are expressed in the use-case view; the design view captures the problem space and solution space; the process view models the systems processes; the implementation view addresses the implementation of the system; and the deployment view.

There are several diagrams providing different viewpoints of the system (e.g., use-case diagrams, class and object diagrams, activity diagrams, sequence diagrams, state diagrams, etc.), and these provide the blueprint of the software. Constraints may be placed on the UML models using a natural language such as English or OCL.

The main advantages of UML are:

Advantages of UML
State of the art visual modeling language with a rich expressive notation.
Study of the proposed system before implementation
Visualization of architecture design of the system.
Mechanism to manage complexity of a large system.
Visualization of system from different viewpoints. The different UML diagrams provide a different view of the system.
Enhanced understanding of implications of user behavior.
Use cases allow a description of typical user behavior.
Use cases provide a mechanism to communicate the proposed behavior of the software system, to describe what it will do, and serves as the basis of development and testing.

Table 10.6. Advantages of UML

11
Technology Transfer

11.1 Introduction

Technology transfer is concerned with the practical exploitation of new technology. The new technology is developed by an academic or industrial research group, and the objective of technology transfer is to facilitate the use of the technology in an industrial environment. The transfer of new technology and leading edge methods to an industrial environment needs to take place in a controlled manner. It cannot be assumed that a new technology or method will necessarily suit an organization, and the initial focus is concerned with piloting the new technology and measuring the benefits gained from its use. The pilot(s) will provide insight into the effectiveness of the new technology as well as identifying areas that require further improvement prior to the general deployment of the technology throughout the company. Feedback from the pilot(s) is provided to the research groups and further improvements and pilots take place as appropriate. In some instances, it may be that commercial exploitation of the technology is inappropriate. This may be due to immaturity of the technology, the fact that the technology does not suit the culture of the company, or the fact that the evaluation of the technology did not achieve the required criteria and is not expected to in the near future.

A pilot of new technology is an experiment to determine the effectiveness of the technology, and to enable an informed decision to be made on the benefits of transferring the new technology throughout the company. The pilot is generally limited to one part of the company or to one specific group in the company. A pilot needs to be planned and this includes deciding which people will participate, the provision of training for the participants, and the identification of objective criteria to evaluate the new technology. The objective criteria needs to be identified prior to the commencement of the pilot, and the results of the pilot are then compared to the evaluation criteria. Further analysis of the evaluation results then takes place. This allows an informed decision to be made as to whether deployment of the technology throughout the entire company is appropriate.

Organization culture needs to be considered for effective technology transfer. Every organization has a distinct culture and this is reflected in the way that people work and in the way in which things are done in the organization. Organization culture includes the ethos and core values of the organization, its commitment or resistance to change, and so on. The transfer of new technologies will be easier in cultures where there is an openness and willingness to change. However, in other cultures the deployment of new technology may be difficult due to resistance to change within the organization.

11.2 Formal Methods and Industry

The formal methods community have developed elegant formalisms to assist the development of high-quality software. However, in practice most software developers find formal methods unusable, whereas most of the formal methods community seem unable to understand why software developers find their notation and methods to be difficult. The chasm between the formal methods community and industrial programmers has been considered in [Web:93] and various characteristics of a good formal method have been outlined.

It is not, of course, the role of the formal methods community to sell formal methods to industry; rather, their role is to develop notations and methods that are usable, provide education on applying formal methods to students and interested industrialists, and to provide expert help during pilots of formal methods in industrial environment. However, in order to develop a usable formal method it is essential that the formal methods community has a better understanding of the needs of industry and the real commercial constraints of industrial projects.

An industrial project consists of a project team, and each team member has various responsibilities in the software development process (e.g., requirements, definition, design, implementation, software testing, configuration management, project management). A project is subject to strict budget, timeliness, and quality constraints, and the project manager is responsible for delivering a high-quality product on time and within budget to the customer. An industrial project follows a defined software process, and there is a need to define the process to be followed when formal methods are part of the process. Formal methods need to be piloted prior to their general deployment in an organization to ensure that there are real benefits gained in higher quality software from their use and that the commercial constraints (e.g., budget, timeliness, and quality) are addressed. Late projects can cause the loss of a market opportunity or cause major customer dissatisfaction and a loss of credibility. Budget overruns and quality problems can lead to financial loss for the company.

Industry is concerned with finding the most cost-effective solution to delivering high-quality software on time and within budget. The natural question [Wic:00] is *Under what circumstances does formal methods provide the most cost effective solution to an industrial problem.*[1] Any selling of formal methods

---

[1] I see the current answer to the question of where formal methods provide a cost effective solution as being applications of formal methods to the regulated sector, as this sector requires certification

to industry must be realistic, as an overselling of the benefits of formal methods has led to a negative perception of the technology.[2] This is since industrialists have experienced difficulties in working with the immaturity of current formal methods, and in practice formal methods have not been used extensively due to problems with their usability.

The development of standards such as 00-55 (British Defense Standard for the procurement of safety critical software), DO-178B (U.S. standard for certification of safety critical avionics software), and IEC 61509 (international standard for critical systems) all mention formal methods as a way to ensure high-quality software. The 00-55 Defense Standard actually mandates[3] the use of formal methods for formal code verification and specification animation.

The safety critical field is one area where the use of formal methods has shown good benefits and where the verification of correctness is essential. Quality and safety cannot be compromised in safety critical systems, and the regulated sector has provided some good case studies in the application of formal methods. These include the Darlington Nuclear generation station in Canada where mathematical techniques were employed to certify the shutdown software of the plant. Other applications include the use of formal methods to certify the safety critical software is the Paris metro signaling system. The regulatory sector is concerned with certification of the code with respect to the requirements, and timeliness is less important than the actual certification. Regulatory systems are generally expensive as there are often several organizations involved, and the verification of the correctness of these products is time consuming.

The application of formal methods to mainstream software engineering has been less successful. Mainstream software engineering is subject to stringent commercial constraints and the experience to date is that the use of formal methods does not provide any appreciable gain in quality, timeliness, or productivity.[4] Time to market is often the main commercial driver in mainstream software engineering.

---

that the code meets stringent safety critical and security critical requirements. The cost of certification in the regulated sector is high but formal methods can be employed to demonstrate to regulators that the code conforms to the requirements. The benefit gained is the extra quality assurance gained as a result of the use of formal methods. The reliability requirements in the regulated sector are much higher than for conventional systems, and the cost of testing may make it infeasible to verify that these reliability requirements have been achieved. This makes the use of formal methods a cost-effective solution. The demonstration includes mathematical proof and testing. As the maturity of formal methods evolves there may be other circumstances in which formal methods provide the most cost-effective solution.

[2] The overselling of formal methods was discussed briefly in Chapter 1 with respect to the formal verification of the VIPER chip.

[3] The new revisions of Defence Standards 00-55 and 00-56 are now less prescriptive.

[4] The CICS project at IBM claimed a 9% increase in productivity. However, Peter Lupton of IBM outlined difficulties that the engineers had with formal specification in Z at the FME'93 conference. Care needs to be taken with measurements in software engineering as some software metrics are unsound. For example, it is easy to increase the productivity of an organization (income per employee) by a redundancy program. Programmer productivity in terms of lines of code per week are also unsound as a measure of productivity as the quality of the code needs to be considered.

11.3 Usability of Formal Methods

There are practical difficulties associated with the usability of formal methods. It seems to be assumed that programmers and even customers are willing to become familiar with the mathematics used in formal methods. There is little evidence to suggest that customers would be prepared to use formal methods.[5] Customers are concerned with their own domain and speak the technical language of that domain.[6] Often, the use of mathematics is an alien activity that bears little resemblance to their normal practical work. Programmers are interested in programming rather than in mathematics, and generally have no interest in becoming mathematicians.[7]

However, the mathematics involved in most formal methods is reasonably elementary, and, in theory, if both customers and programmers are willing to learn the formal mathematical notation, then a rigorous validation of the formal specification can take place to verify its correctness. Both parties can review the formal specification to verify its correctness, and the code can be verified to be correct with respect to the formal specification. It is usually possible to get a developer to learn a formal method, as a programmer has some experience of mathematics and logic; however, in practice, it is more difficult to get a customer to learn a formal method.

This means that often a formal specification of the requirements and an informal definition of the requirements using a natural language are maintained. It is essential that both of these documents are consistent and that there is a rigorous validation of the formal specification. Otherwise, if the programmer proves the correctness of the code with respect to the formal specification, and the formal specification is incorrect, then the formal development of the software is incorrect. There are several techniques to validate a formal specification and these are described in [Wic:00]:

Technique	Description
Proof	This involves demonstrating that the formal specification satisfies key properties of the requirements. The implementation of the software will also need to preserve these properties. The proof will generally employ rigorous mathematical reasoning.

---

[5] The domain in which the software is being used will influence the willingness of the customers to become familiar with the mathematics required. Certainly, in mainstream software engineering I have not detected any interest from customers and the perception is that formal methods are unusable; however, in some domains such as the regulated sector there is a greater willingness of customers to become familiar with the mathematical notation.

[6] My experience is that most customers have a very limited interest and even less willingness to use mathematics. There are exceptions to this especially in the regulated sector.

[7] Mathematics that is potentially useful to software engineers is discussed in Chapter 2.

Software inspections	This involves a Fagan like inspection to perform the validation. It may involve comparing an informal set of requirements (unless the customer has learned the formal method) with the formal specification.
Specification animation	This involves program (or specification) execution as a way to validate the formal specification. It is similar to testing.
Tools	Tools provide some limited support in validating a formal specification.

Table 11.1. Techniques for Validation of Formal Specification

Why Are Formal Methods Difficult?

Formal methods are perceived as being difficult to use and of offering limited value in mainstream software engineering. Programmers receive some training in mathematics as part of their education. However, in practice, most programmers who learn formal methods at university never use formal methods again once they take an industrial position.

It may well be that the very nature of formal methods is such that it is suited only for specialists with a strong background in mathematics. Some of the reasons why formal methods are perceived as being difficult are:

Factor	Description
Notation / intuition	The notation employed differs from that employed in mathematics. Intuition varies from person to person. Many programmers find the notation in formal methods to be unintuitive.
Formal specification	It is easier to read a formal specification than to write one.
Validation of formal Specification	The validation of a formal specification using proof techniques or a Fagan-like inspection is difficult.
Refinement[8]	The refinement of a formal specification into successively more concrete specifications with proof of validity of each refinement step is difficult and time consuming.
Proof	Proof can be difficult and time consuming.
Tool support	Many of the existing tools are difficult to use.

Table 11.2. Factors in Difficulty of Formal Methods

---

[8] I doubt that refinement is cost effective for mainstream software engineering. However, it may be useful in the regulated environment.

Characteristics of a Usable Formal Method

It is important to investigate ways by which formal methods can be made more usable to software engineers. This may involve designing more usable notations and better tools to support the process. Practical training and coaching to employees can help also. Some of the characteristics of a usable formal method are the following:

Characteristic	Description
Intuitive	A good intuitive notation has potential as a usable formal method. Intuition does vary among people.
Teachable	A formal method needs to be teachable to the average software engineer. The training should include (at least) writing practical formal specifications.
Tool support	Usable tools to support formal specification, validation, refinement, and proof are required.
Adaptable to change	Change is common in a software engineering environment. A usable formal method should be adaptable to change.
Technology transfer path	The process for software development needs to be defined to include formal methods. The migration to formal methods needs to be managed.
Cost[9]	The use of formal methods should be a cost effective (timeliness, quality, and productivity). There should be a return on investment from their use.

Table 11.3. Characteristics of a Usable Formal Method

11.4 Pilot of Formal Methods

The transfer of new technology to the organization involves a structured pilot of the new technology using objective evaluation criteria. A decision is made following the pilot to either conduct further pilots, abandon the technology, or transfer the technology within the company. The steps for a pilot of formal methods are:

---

[9] A commercial company will expect a return on investment from the use of a new technology. This may be reduced software development costs, improved quality, improved timeliness of projects, or improvements in productivity. A company does not go to the trouble of deploying a new technology just to satisfy academic interest.

Step	Description
Overview of technology	This provides the motivation for technology transfer. An organization (or group) receives an overview of a new technology that offers potential. E.g., this may be an approach such as Z or VDM.
Identify pilot project	The technology may be sufficiently promising for the organization to consider a pilot. This involves identifying a suitable project for the pilot and the identification of the project participants.
Process for pilot	Define the project's software process to be followed for the pilot. The process will detail where formal methods will be used in the lifecycle.
Training	Provide training on the new technology (formal method) and the process for the pilot. The training will require the students to write formal specifications.
Evaluation criteria (pilot)	Identify objective criteria to judge the effectiveness of the new technology. This includes gathering data for: • Productivity measurements • Quality measurements • Timeliness measurements
Support (pilot)	Provide on-site support to assist the developers in preparing formal specifications. This may require consultants.
Conduct pilot	The pilot is conducted and the coordinator for the pilot will work to address any issues that arise. Data are gathered to enable objective evaluation to take place.
Postmortem[10]	A postmortem is conducted after the pilot to consider what went well and what did not. The evaluation criteria are compared against the gathered data and recommendations are made to either conduct further pilots, abandon the technology, or institutionalize the new technology.

Table 11.4. Steps for Pilot of Formal Methods

---

[10] It may well be that the result of a pilot of formal methods results in a decision that the methodology is inappropriate for the company at this time. The bottom line is whether formal methods provide a more cost effective solution to software engineering problems that other engineering approaches. Further pilots may be required before a final decision can be made.

11.5 Technology Transfer of Formal Methods

The transfer of new technology to the organization is done in a controlled manner. The steps in the technology transfer are:

Step	Description
Decision to institutionalize	A decision is made to transfer the new technology throughout the organization. The results of the pilot justify the decision.
Software development Process	Update the software development process to define how formal methods are used as part of the development process.
Training	Provide practical training to all affected staff in the company. The training will include writing formal specifications.
Audits	Verify that the new process is being followed and that it is effective by conducting audits. The results of the audits are reported in management.
Improvements	Potential improvements to the technology or process are identified and acted upon.

Table 11.5. Steps for Technology Transfer of Formal Methods

11.6 Summary

Technology transfer is the disciplined transfer of new technology to a company and is concerned with the practical exploitation of the new technology. It cannot be assumed that a new technology or method will necessarily benefit an organization, and the initial focus is concerned with piloting the new technology and measuring the benefits gained from its use. The pilot(s) will provide insight into the effectiveness of the new technology as well as identifying areas that require further improvement prior to deployment of the technology throughout the company.

The pilot is generally limited to one part of the company or to one specific group in the company. A pilot needs to be planned and this includes deciding which people will participate, the provision of training for the participants, and the identification of objective criteria to evaluate the new technology. The objective criteria needs to be identified prior to the commencement of the pilot, and the results of the pilot are then compared to the evaluation criteria and further analysis and a postmortem take place. This allows an informed decision to be made as to whether the deployment of the new technology throughout the entire company is appropriate.

References

Ack:94 Aristotle the Philosopher. J.L. Ackrill. Clarendon Press, Oxford. 1994.

Ada:84 Optimizing Preventive Service of Software Products. E. Adams. IBM Research Journal, 28(1), pp. 2-14, 1984.

Bec:00 Extreme Programming Explained. Embrace Change. Kent Beck. Addison Wesley. 2000.

BeM:86 Software Reliability. State of the Art Report. Edited by A. Bendell and P. Mellor. Pergamon Infotech. 1986.

BjJ:78 The Vienna Development Method. The Meta language. Dines Bjorner and Cliff Jones. *Lecture Notes in Computer Science* (61). Springer Verlag. 1978.

BjJ:82 Formal Specification and Software Development. Dines Bjorner and Cliff Jones. Prentice Hall International Series in Computer Science. 1982.

Boa:66 Mathematical Methods in the Physical Sciences. M.L. Boas. Wiley International Series. 1966.

Boe:81 Software Engineering Economics. Barry Boehm. Prentice Hall. 1981.

Boe:88 A Spiral Model for Software Development and Enhancement. Barry Boehm. Computer. May, 1988.

Bou:94 Formalization of Properties for Feature Interaction Detection. Wiet Bouma et al. IS&N Conference. Springer Verlag. 1994.

Brk:75 The Mythical Man Month. Fred Brooks. Addison Wesley. 1975.

Brk:86 No Silver Bullet. Essence and Accidents of Software Engineering. Fred Brooks. Information Processing. Elsevier. Amsterdam, 1986.

Bro:90 Rational for the Development of the U.K. Defence Standards for Safety Critical software. Compass Conference. 1990.

But:99 The VDM♣ Reference. Andrew Butterfield. Foundations and Methods Group Technical Report. University of Dublin. Trinity College Dublin. 1999.

But:00 VDM♣ Mathematical Structures for Formal Methods. Presentation by Andrew Butterfield. Foundations and Methods Group. Trinity College, Dublin. 19th May 2000.

ClN:02 Software Product Lines. Practices and Patterns. Paul Clements and Linda Northrop. Addison-Wesley. 2002.

CKS:03 CMMI. Guidelines for Process Integration and Product Improvement. Mary Beth Chrissis, Mike Conrad, and Sandy Shrum. Addison-Wesley. 2003.

CoM:90 Engineering Software under Statistical Quality Control. Richard H. Cobb and Harlan D. Mills. IEEE Software. 1990.

CSE:00 Unified Modeling Language. Technical Briefing No. 8. Centre for Software Engineering. Dublin City University. Ireland. April 2000.

Dij:72 Structured Programming. E.W. Dijkstra. Academic Press. 1972.

Dil:90 Z. An Introduction to Formal Methods. Antoni Diller. John Wiley and Sons. England. 1990.

Fag:76 Design and Code Inspections to Reduce Errors in Software Development. Michael Fagan. IBM Systems Journal 15(3). 1976.

Fen:00 Quantitative Analysis of Faults and Failures in a Complex Software System. Norman Fenton and Niclas Ohlsson. IEEE Transactions on Software Engineering, 26(8), pp. 797-814. 2000.

Fly:67 Assigning Meaning to Programs. R. Floyd. Mathematical Aspects of Computer Science. American Mathematical Society. 1967.

Geo:91 The RAISE Specification Language. A Tutorial. Chris George. Lecture Notes in Computer Science (552). Springer Verlag. 1991.

Ger:94   Experience with Formal Methods in Critical Systems. Susan Gerhart, Dan Craighen, and Ted Ralston. IEEE Software. January 1994.

Glb:94   Software Inspections. Tom Gilb and Dorothy Graham. Addison Wesley. 1994.

Gri:81   The Science of Programming. David Gries. Springer Verlag. Berlin. 1981.

HB:95   Applications of Formal Methods. Edited by Michael Hinchey and Jonathan Bowen. Prentice Hall International Series in Computer Science. 1995.

Her:75   Topics in Algebra. I.N. Herstein. 2nd Edition. John Wiley and Co. 1975.

Hey:66   Intuitionism (An Introduction). A. Heyting. North Holland Publishing Company. 1966.

Hor:69   An Axiomatic Basis for Computer Programming. C.A.R. Hoare. Communications of the ACM. Volume 12, No. 10. 1969.

Hor:85   Communicating Sequential Processes. C.A.R. Hoare. Prentice Hall International Series in Computer Science. 1985.

Hoa:95   Application of the B-Method to CICS. Jonathon P. Hoare. In Applications of Formal Methods. Editors Michael Hinchey and Jonathon P. Bowen. Prentice Hall International Series in Computer Science. 1995.

InA:91   Practical Formal Methods with VDM. Darrell Ince and Derek Andrews. McGraw Hill International Series in Software Engineering. 1991.

Jac:99a   The Unified Software Modeling Language User Guide. Ivar Jacobson, Grady Booch, and James Rumbaugh. Addison-Wesley. 1999.

Jac:99b   The Unified Software Development Process. Ivar Jacobson, Grady Booch, and James Rumbaugh. Addison-Wesley. 1999.

Jan:97   On a Formal Semantics of Tabular Expressions. R. Janicki. Technical Report CRL 355. Communications Research Laboratory. McMaster University. Ontario. 1997.

Jon:90   Systematic Software Development Using VDM. Cliff Jones. Prentice Hall International Series in Computer Science. 2nd Edition. 1990.

Kel:97   The Essence of Logic. John Kelly. Prentice Hall. 1997.

Kuh:70   The Structure of Scientific Revolutions. Thomas Kuhn. University of Chicago Press. 1970.

Lak:76   Proof and Refutations. The Logic of Mathematical Discovery. Imre Lakatos. Cambridge University Press. 1976.

Let:03   How Software Engineers Use Documentation. Timothy Lethbridge et al. IEEE Software. November. 2003.

LiT:96   Cleanroom Software Engineering Reference Model. R. Linger and C. Trammell. Software Engineering Institute. Carnegie Mellon. 1996.

Lof:84   Intuitionist Type Theory. Martin Lof. Study in Proof Theory Series. Lecture Notes. Bibliopolis. Naples. 1984.

Mac:90   Conceptual Models and Computing. PhD Thesis. Micheal Mac An Airchinnigh. Department of Computer Science. University of Dublin. Trinity College, Dublin. 1990.

Mac:93   Formal Methods and Testing. Micheal Mac An Airchinnigh. Tutorials of the 6th International Software Quality Week. Software Research Institute. San Francisco. 1993.

McD:94 MSc. Thesis. Eoin McDonnell. Department of Computer Science. Trinity College Dublin. 1994.

Men:87   Introduction to Mathematical Logic. Elliot Mendelson. Wadsworth and Cole/Brook, Advanced Books & Software. 1987.

Mil:79   Structured Programming: Theory and Practice. H. Mills, R. Linger and B. Witt. Addison-Wesley. 1979.

Mil:87   Cleanroom Software Engineering. H. Mills, M. Dyer, R. Linger. IEEE Software. Vol. 4. 1987.

Mil:89   Communication and Concurrency. Robin Milner. International Series in Computer Science. Prentice Hall. 1989.

MOD:91a Def Stan 00-55 (Part 1). Requirements for Safety Critical Software in Defence Equipment. Interim Defence Standards. U.K. 1991

MOD:91b Def Stan 00-56. Guidance for Safety Critical Software in Defence Equipment. Interim Defence Standards. U.K. 1991.

ORg:97   Applying Formal Methods to Model Organizations and Structures in the Real World. PhD Thesis. Department of Computer Science. Trinity College, Dublin. 1997.

ORg:02   A Practical Approach to Software Quality. Springer Verlag. 2002.

Par:01   Software Fundamentals. Collected Papers by David L. Parnas. Edited by Danield Hoffman and David Weiss. Addison Wesley. 2001.

Par:72   On the Criteria to be used in Decomposing Systems into Modules. David L. Parnas. Communications of the ACM, 15(12). 1972.

Par:92   Tabular Representation of Relations. David L. Parnas. CRL Report 260. McMaster University, Canada. 1992.

Par:93   Predicate Calculus for Software Engineering. David L.Parnas. IEEE Transactions on Software Engineering, 19(9). 1993.

Par:94   Inspection of Safety Critical Software Using Function Tables. David L. Parnas. Proceedings IFIP 13th World Computer Congress, vol. 3, North Holland, pp. 270-277. 1994.

Par:95   Functional Documents for Computer Systems. David L. Parnas and J. Madley. Science of Computer Programming. Elsevier. 1995.

Pau:93   Key Practices of the Capability Maturity Model. Software Engineering Institute. Mark Paulk et al. 1993.

PaW:01   Active Design Reviews: Principles and Practice. David L. Parnas and David M. Weiss. In Software Fundamentals. Collected Papers by David L. Parnas. Edited by Daniel Hoffman and David Weiss. Addison Wesley. 2001.

Pif:91   Discrete Mathematics. An Introduction for Software Engineers. Mike Piff. Cambridge University Press. 1991.

Pol:57   How to Solve It. A New Aspect of Mathematical Method. Georges Polya. Princeton University Press. 1957.

Pop:97   The Single Transferable Voting System. Michael Poppleton. IWFM'97. Editors Gerard O'Regan and Sharon Flynn. Springer Verlag. 1997.

Pro:99   Cleanroom Software Engineering. Technology and Process. S. Prowell, C. Trammell, R. Linger, and J. Poore. SEI Series in Software Engineering. Addison-Wesley. 1999.

Ros:87   Introduction to Probability and Statistics for Engineers and Scientists. Sheldon M. Ross. Wiley Publications. 1987.

Roy:70   The Software Lifecycle Model (Waterfall). W. Royce. Proceedings of Westcon. August 1970.

Spi:92   The Z Notation. A Reference Manual. J. M. Spivey. Prentice Hall International Series in Computer Science. 1992.

Sta:92   Practical Experiences with Safety Assessment of a System for Automatic Train Control. Proceedings of SAFECOMP'92, Zurich, Switzerland. Pergamon Press, Oxford, UK. 1992.

Tie:91   The Evolution of Def Stan 00-55 and 00-56. An Intensification of the Formal Methods Debate in the UK. Margaret Tierney. Research Centre for Social Sciences. University of Edinburgh. 1991.

Web:93   Selling Formal Methods to Industry. Debora Weber-Wulff. FME'93. LNCS 670. 1993.

Wic:00   A Personal View of Formal Methods. B. A. Wichmann. National Physical Laboratory. March 2000.

Wrd:92   Formal Methods with Z. A Practical Approach to Formal Methods in Engineering. J. B. Wordsworth. Addison Wesley. 1992.

Abbreviations

ACM	Associated Computer Machinery
AMN	Abstract Machine Notation
ATM	Automated Teller Machine
CCS	Calculus Communicating Systems
CICS	Customer Information Control System
CMM®	Capability Maturity Model
CMMI®	Capability Maturity Model Integration
CSP	Communicating Sequential Processes
DAG	Directed Acyclic Graph
DOD	Department of Defense
ESI	European Software Institute
FSM	Finite State Machine
ICSE	International Conference Software Engineering
IEEE	Institute of Electrical Electronic Engineers
IFIP	International Federation for Information Processing
ISO	International Standards Organization
LD	Limited Domain (relation)
LPF	Logic Partial Functions
MTBF	Mean Time Between Failure
MTTF	Mean Time To Failure
NRL	Naval Research Laboratory
OCL	Object Constraint Language

OMG	Object Management Group
OMT	Object management technique
OOSE	Object-Oriented Software engineering
PIN	Personal Identification Number
PVS	Prototype Verification System
RACE	Research Advanced Communications Europe
RAISE	Rigorous Approach to Industrial Software Engineering
RUP	Rational Unified Process
SCORE	Service Creation Object Reuse Environment
SDI	Strategic Defence Initiative
SDL	Specification and Descriptive Language
SEI	Software Engineering Institute
SPI	Software Process Improvement
SPICE	Software Process Improvement and Capability determination
SQA	Software Quality Assurance
SSADM	Structured Systems Analysis and Design Method
TAM	Trace Assertion Method
UML	Unified Modeling Language
VDL	Vienna Definition Language
VDM	Vienna Development Method
VDM-SL	VDM Specification Language
XP	Extreme Programming

Index